Cover Illustration: The Hebrew Candlestick in the Tabernacle
Courtesy of Gerth Median Gmbh; Postfach 1148; 3567; Asslar, Germany

(Note: *The Tabernacle of God in the Wilderness* illustration book is currently out of print)

****

The Articles of Configuration
THE GENESIS PROJECT

Carl D. Armstrong

AuthorHouse™
1663 Liberty Drive
Bloomington, IN 47403
www.authorhouse.com
Phone: 1-800-839-8640

©2010 Carl D. Armstrong. All rights reserved.

No part of this book may be reproduced, stored in a retrieval system, or transmitted by any means without the written permission of the author.

However, permission is granted to newspapers and periodical publications to serially print the articles contained in each chapter of this book. Only one of the articles may be printed in any given edition and The Articles of Configuration book must be credited as the source for the printed article.

First published by AuthorHouse   10/18/2010

ISBN: 978-1-4520-5457-5 (sc)
ISBN: 978-1-4520-5459-9 (e)
ISBN: 978-1-4520-5458-2 (hc)

Library of Congress Control Number: 2010911312

Printed in the United States of America

This book is printed on acid-free paper.

Contents

Chain of Custody of Genesis 1
Earth's Magnetic Shield In The Bible? 11
Are You Smarter Than a PhD Theoretical Physicist? 19
The Light At The Top Of The Mountain 29
Expanding Universe In Scriptures? 39
The Lord's Time Travel Paradox 49
Do Life Forms Exist Among The Stars? 57
If I Try And Bend That Far, I'll Break 65
MRI Of Brain Seen By Hebrew Prophets? 77
Unseen Dimensions Life Source? 87
The Articles Of Configuration 96

DEDICATION

It is the glory of God to conceal a thing: but the honour of kings is to search out a matter.

And whosoever shall exalt himself shall be abased; and he that shall humble himself shall be exalted.

This series of articles is dedicated to all the servants of mankind who have searched out both scientific and spiritual truths. For truth has not always been received with open arms. Rather it has often been greeted with derision and persecution. To these servants, both living and dead, we owe a great debt.

Personally, this author owes a great debt to the many tutors and governors that he has encountered over the years. And specifically, gratitude is expressed toward my wife and children for graciously accommodating the time I have spent in studying scientific and spiritual truths.

And also thanks to some very, very dear brothers and sisters in the Lord—without their encouragement, this effort would not have gone forward.

Carl Armstrong

Foreword

Science and theology has been locked together in a mighty struggle of ideas. The process, at times, has been ugly as a caterpillar chomping on leaves, regurgitating and trying to build a cocoon of how the universe is really constituted. But this cocoon, like the human brain, has an inside tension between left brained logic and right brained dreaming. A mighty struggle ensues and at last a butterfly emerges from the cocoon and its glistening wings dry in the sun.

As the butterfly discovers its destiny and ascends in flight into the brilliant blue sky filled with fluffy white clouds, a serene revelation of peace envelops the butterfly. It realizes that its one wing of science and its other wing of theology are working in harmony under a Master control. Thrilled and awed, the butterfly uses both wings to ascend far above worldly ideas into realms that it once only dreamed of.

Carl D. Armstrong

Note: Each chapter of the book contains an individual—newspaper type article that has been combined with other stand—alone chapter articles to make a serial publication.

Author Carl Armstrong is an engineering graduate of the University of Missouri School of Mines and Metallurgy. He worked with major US companies for a number of years and has now retired.

Article One

CHAIN OF CUSTODY OF GENESIS

What Happened Before Genesis?

In the beginning God created the heaven and the earth. These are the opening words of the Bible—but what happened before the heaven and earth were created? From what were they created and how did they come to be? Current scientific theory and observations suggest that the universe is about 13.7 billion years old and that the earth is about 4.5 billion years old. Radiometric age dating of the oldest known earth materials, meteorite materials, and lunar materials have placed the earth age at about 4.5 billion years and the sun's age being slightly older at about 4.8 billion years.

So, which is it for our heaven and earth? The Bible tells us that a day with the Lord—instead of 24 hours—could be equivalent to one thousand years. So, if we take 1000 years for each of the seven days of creation and add our recorded history time elapse of about 6,000 years, then we could even propose a max earth age of around 13,000 years. This is a far cry from the about 4.5 billion years that the scientists are suggesting. What a paradox! Is the Bible wrong, or—are scientists wrong in assuming that radiometric dating applies? Or, could both be right? Apparently the jury is still debating this paradox?

In a court case, there is a concept called 'chain of custody' which describes the process for insuring that evidence from the original scene remains secure so that the facts can be truthfully presented before a jury. So, science has its samples and methods to present as evidence. But, what about the Bible? Before Genesis chapter one, verse one, we have no information to examine—we are stuck! Not so, we just haven't known where to look for the pre-Genesis information!

Most everyone is familiar with the scripture: *And the LORD God formed man of the dust of the ground, and breathed into his nostrils the breath of life; and man became a living soul.* So, if man was created from *the dust of the ground*, where did this dust come from? Yes, we know that dust comes from the earth, but where did the dust in the earth come from?

Now, we are ready to present our chain of custody and it comes from an unlikely place—from near the middle of the Bible. It is presented as a flashback in Solomon's book of wisdom — Proverbs. And it is narrated by *wisdom* who gives an account of the creation process which began with *the dust of the world* (cosmos).

PROVERBS 8:1 ... 31 *Doth not wisdom cry? and understanding put forth her voice? She standeth in the top of high places, by the way in the places of the paths. ...*

The LORD possessed me in the beginning of his way, before his works of old. I was set up from everlasting, **from the beginning, or ever the earth was.** *When there were no depths, I was brought forth; when there were no fountains abounding with water.*

Before the mountains were settled, before the hills was I brought forth: **While as yet he had not made the earth** *nor the fields, nor* <u>**the highest part of the dust of the world**</u>**.** *When he prepared the heavens, I was there: when he set a compass upon the face of the depth:*

When he established the clouds above: when he strengthened the fountains of the deep: When he gave to the sea his decree, that the waters should not pass his commandment: when he appointed the foundations of the earth: Then I was by him, as one brought up with him: and I was daily his delight, rejoicing always before him; Rejoicing in the habitable part of his earth; and my delights were with the sons of men.

We are presented with an 'eye witness' account of a process that began before *as yet he had not made the earth*—but it doesn't end there. The *highest part of the dust of the world* would seem to refer to a time when the itty bitty dust-like elements were formed within the stars of the galaxies — the likely result of super nova explosions. And then, we have the formation of these elements into dust particles agglomerating into asteroid clouds around a central star. The asteroids agglomerate into roughly shaped planets which existed *before the mountains were settled*.

As multi-spectrum optics became available to peer into the depths of our Milky Way Galaxy, large clouds of dust and gases have been observed. It has been estimated that about 15% of the galactic disc consists of dust and gases. Scientists have observed that the constellation Orion is a star 'birthing' nursery and our sun is in the Orion spur of the Milky Way. *Seek him that maketh the seven stars and* **Orion**, *and turneth the shadow of death into the morning, and maketh the day dark with night: that calleth for the waters of the sea ...*

And then, where did the water come from? Did not the element of water come from the ice in the asteroid belts *when he set a compass upon the face of the depth:* (Sorry, to disappoint anyone in the flat earth society, but a compass does draw a circle and the deep oceans do have a spherical face)?

Now that we have established the scriptural chain of custody from *the highest part of the dust of the world* to the beginning verse of Genesis, what conclusion can we draw about the age of the elements in the earth? What was the time period *from everlasting, from the beginning, or ever the earth was?* The scriptures don't say—it could have been 4.5 billion years, 4.5 trillion years—or perhaps e*verlasting* is completely outside of time and space. Therefore, the process of creating the elements that were used to create *the heaven and the earth* in its configuration would seem to be independent of the age of elements used to assemble it. Now, our scientists can go busily about their work of determining the age of the elements *from the dust of the world.*

But, we feel a tap on our collective shoulder from ardent scientists and the question is posed, "You forget one little detail – don't the scriptures say that the sun was created on the fourth day? This flies in the face of what astronomers observe about planetary formation around a central star. How can we trust the scriptures, if they get the sequence wrong?"

A good question, we must admit. No doubt many scientists and theologians have struggled with this over the centuries. First, we will look at the pertinent scriptures involved from Genesis One.

GENESIS *1:1 In the beginning God created the heaven and the earth.*

2 And the earth was without form, and void; and darkness was upon the face of the deep. And the Spirit of God moved upon the face of the waters.

3 And God said, Let there be light: and there was light. ...

4 And God saw the light, that it was good: and God **divided the light from the darkness**.

5 And God called the light Day, and the darkness he called Night. And **the evening and the morning were the first day.** ...

13 And the evening and the morning were **the third day**.

*14 And God said, Let there be lights in the firmament of the heaven **to divide the day from the night**; and let them be for signs, and for seasons, and for days, and years:*

15 And let them be for lights in the firmament of the heaven to give light upon the earth: and it was so.

16 And God <u>made two great lights</u>; the greater light to rule the day, and the lesser light to rule the night: he made the stars also.

17 And God set them in the firmament of the heaven to give light upon the earth,

18 And to rule over the day and over the night, and to divide the light from the darkness: and God saw that it was good.

*19 And the evening and the morning were **<u>the fourth day</u>**.*

So, it seems like an open and shut case. If we believe the sun was first formed, then the planets were formed from the space debris orbiting around it, so the Bible must be wrong. Or, is it?

To delve into this, we must ask some very pertinent questions. Just what is happening in verse 14 when the night is divided from the day? Does this have something to do with the rotation of the earth on its axis? How are the verse 14 signs and seasons brought about? Does this have something to do with the tilt of the axis of the earth as it orbits around the sun? If the sun were created first, why would there be the verse 1 darkness on the earth on the first day?

To begin to answer these questions, we will visit other 'flashbacks' to Genesis and before which are given in the 26th and 38th chapters of Job. These chapters give us information on how the earth was assembled together and nursed into being habitable.

JOB 38:1 Then the LORD answered Job out of the whirlwind, and said, 2 Who is this that darkeneth counsel by words without knowledge? 3 Gird up now thy loins like a man; for I will demand of thee, and answer thou me.

*4 **Where wast thou when I laid the foundations of the earth?** declare, if thou hast understanding. 5 Who hath laid the measures thereof, if thou knowest? or who hath stretched the line upon it? 6 Whereupon are the foundations thereof fastened? or who laid the corner stone thereof; 7 When the morning stars sang together, and all the sons of God shouted for joy? 8 Or who shut up the sea with doors, when it brake forth, as if it had issued out of the womb? 9 **When I made the cloud the garment thereof, and thick darkness a swaddlingband for it,***

When we lay the foundations of a house, we use construction elements that have been preassembled for building the house. For example, if redwood from a thousand year old tree is used, it doesn't means that the house itself is one thousand years old, it simply means that its construction includes one thousand year old plus members. And, just in case anyone from the flat earth society is reading this, it is instructive to read Job's pre-Galileo and pre-Copernicus description of what a modern day NASA astronaut might have seen in observing the earth from afar. Who can forget NASA beaming the Christmas Eve pictures of the earth hanging *upon nothing* and the division of the day and night on the *compassed* surface of the earth? Note that at first there are **thick clouds** and then he **_spreadeth_** his cloud until **day and night come to an end**.

JOB 26:7 *He stretcheth out the north over the empty place,* **and hangeth the earth upon nothing**. *8 He bindeth up the waters in his* **thick clouds***; and the cloud is not rent under them. 9 He holdeth back the face of his throne,* **and spreadeth his cloud upon it. 10 He hath compassed the waters with bounds, until the day and night come to an end.**

Now, that we have visited the flashbacks regarding creation from Job, we will take another look at the first verses of Genesis.

GENESIS 1: *In the beginning God created the heaven and the earth. 2 And the earth was without form, and void;* **and darkness was upon the face of the deep**. *And the Spirit of God moved upon the face of the waters. 3 And God said, Let there be light: and there was light. 4 And God saw the light, that it was good: and* **God divided the light from the darkness,**. *And God called* **the light Day, and the darkness he called Night**. *And the evening and the morning were the first day.*

So, let's summarize a scenario of how this process came about by using our flashback information and the first verses of Genesis. Initially, **darkness was upon the face of the deep** and **thick clouds** are the material used when the LORD said, **I made the cloud the garment thereof, and thick darkness a swaddlingband for it ...**" Then, **God divided the light from the darkness** and the evening of darkness and the light of day resulted in the measuring unit we call a 'day'.

If you lived under the thick shroud of clouds on Venus and someone asked you to choose a lifetime of 100 Venus years or 100 Venus days—

which one would you pick? Strangely enough 100 Venus days would be equivalent to 24,300 earth days or 66.5 earth years. On the other hand, 100 Venus years would be equivalent to 22,500 earth days or 61.5 earth years. One daily rotation of Venus takes longer than its time to complete an orbit around the sun. Its thick *swaddling band* of clouds diffuses the light in such a way that even the 'dark side' has a soft luminance or ashen tint. The rotation of Venus is opposite that of the earth and most other planets making the sun come up in the west and set in the east – some think a chaotic collision caused it to have a retrograde rotation.

Why are we flying off on a tangent to consider the parameters of Venus? The Apostle Paul wrote these words: *Because that which may be known of God is manifest in them; for God hath shewed it unto them. For the invisible things of him from the creation of the world are clearly seen, being understood by the things that are made, even his eternal power and Godhead;*

Perhaps what we are seeing on Venus today parallels what occurred in earth's ancient past. First there is *darkness on the face of the deep* => a *swaddling band* => earth begins rotating so *day and night come to an end* => *the evening and the morning were the first day* =>God *spreadeth his cloud* => sun, moon, and stars become visible from earth => sun is made to *rule the day* and moon is made to *rule the night*.

A more urgent tap on our collective shoulder comes from ardent scientists saying, "By agreeing that the sun was created first and then the earth, you have proved the scriptures wrong. For they plainly say the sun, moon, and stars were created on the fourth day!"

Ooh, this is a big problem—or is it? It is understandable how one might arrive at that conclusion. Let's go back to our source scriptures and look carefully and we will add emphasis by examining the Hebrew source words for CREATED [*bara'*] and MADE [*'asah*].

GENESIS *1:1 In the beginning God* **CREATED** *[bara'] the heaven and the earth.*

2 And the earth was without form, and void; and darkness was upon the face of the deep. And the Spirit of God moved upon the face of the waters.

3 And God said, Let there be light: and there was light. ...

8 And God called the firmament Heaven. And the evening and the morning were the second day. ...

13 And the evening and the morning were **the third day**.

*14 And God said, Let there be lights in the firmament of the heaven to divide the day from the night; and let them be **for signs, and for seasons, and for days, and years**:*

15 And let them be for lights in the firmament of the heaven to give light upon the earth: and it was so.

*16 And God **MADE** ['asah] two great lights; the greater light **TO RULE** the day, and the lesser light **TO RULE** the night: he made the stars also.*

17 And God set them in the firmament of the heaven to give light upon the earth,

18 And to rule over the day and over the night, and to divide the light from the darkness: and God saw that it was good.

*19 And the evening and the morning were **the fourth day**.*

The scripture unambiguously says in verse 1 that God **CREATED** and 'created' is translated from the Hebrew word: *bara'*. However, in verse 16, it says that God **MADE** and this is translated from a different Hebrew word: *'asah*. But, we have a question. In verse 16, did God make the sun, moon, and stars – **or did he make them to do something**. What was the 'something' that he **MADE** them to do? God **MADE** the sun **TO RULE** the day. He **MADE** the moon **TO RULE** the night. How did he bring about this rule – by fine tuning cloud dispersion, rotation of the earth, and orbital tilt of the earth *for signs, and for seasons, and for days, and years?*

For example, when we say, "My boss made me work overtime." – does it mean that he created me? Or, does it mean my boss made me do something? The distinction is very clear! And so it is with the sun and moon – *For the invisible things of him from the creation of the world are clearly seen, being understood by the things that are made ...* We should check out the visible things first in order to understand what—in our limited understanding—is truly the miraculous invisible. St. Augustine once phrased it this way, "that in the first founding of the order of nature, we must not look for miracles, but for what is in accordance with nature."

In the flashback scripture from Proverbs, there is a curious statement about a portion of the earth being habitable. We might conclude from this that some portions of the earth were not yet habitable and the seeding of the earth began like in a garden and eventually filled the whole earth.

PROVERBS 8:28 *When he established the clouds above: when he strengthened the fountains of the deep: 29 When he gave to the sea his decree, that the waters should not pass his commandment: when he appointed the foundations of the earth: 30 Then I was by him, as one brought up with him:*

and I was daily his delight, rejoicing always before him; 31 Rejoicing in **<u>the habitable part</u>** *of his earth; and my delights were with the sons of men.*

GENESIS 2:7 *And the LORD God formed man of* **<u>the dust</u>** *of the ground, and breathed into his nostrils the breath of life; and man became a living soul. 8 And the LORD God* **<u>planted a garden</u>** *eastward in Eden; and there he put the man whom he had formed.*

And of course, the scriptures show the same pattern when Noah took the precious seed of living organisms into the ark with him before the flood—in order to reestablish a colony after the flood. But where did the concept of seed first show up in the Bible?

GENESIS 1:10 *And God called the dry land Earth; and the gathering together of the waters called he Seas: and God saw that it was good. 11 And God said, Let the earth bring forth grass,* **the herb yielding seed**, *and the fruit tree yielding fruit after his kind,* **whose seed is in itself**, *upon the earth: and it was so.*

In order to establish an ecosystem suitable for man, it was necessary to get the earth out of the mud by having grass, herbs, and trees. Where did this seed come from and what does the scripture mean when it says *the seed is in itself*? Did the seed come spontaneously out of the mud and water, or was it already established with life in itself before it was planted in the earth? Like *the highest part of the dust of the world*, could the pattern for forming the seed have come from somewhere else in the universe? Was the pattern created somewhere else over a period of time in the ancient, ancient past? If so, we don't know how long it took to develop the pattern.

Where could the seed pattern have come from? With advances in astronomy, scientists are discovering that there are trillions and trillions of stars and have even calculated that there are more stars than all the sands of the seas of earth. And the probabilities show that it is likely that life exists in other parts of the universe. Given this background the words of the Apostle Paul have special meaning in describing that the invisible ingredient of seed is faith.

HEBREWS 11:3 Through **faith** we understand that **<u>the worlds were framed by the word of God, so that things which are seen were not made of things which do appear</u>**. ...

10 *For he looked for* **a city which hath foundations**, *whose builder and maker is God. 11 Through faith also Sara herself received strength to conceive* **seed**, *and was delivered of a child when she was past age, because she judged him faithful who had promised. 12 Therefore sprang there even of one, and him as good as dead,* **so many as the stars of the sky in multitude, and as the sand which is by the sea shore innumerable**. *13 These all died in faith, not having received the promises,* **but having seen them afar off**, *and were persuaded of them, and embraced them, and* **confessed that they were strangers and pilgrims on the earth**.

If these were strangers and pilgrims on earth, when and where did they come from? Scientists were so excited when a meteorite that was thought to have come from Mars seemed to have a fossilized formerly living organism within it. Most scientists would readily admit that they do not know whether earth's living seed originated on earth or came from somewhere else.

When we look at the flashback from Job chapter 38, it does appear that the spirits of all *the sons of God* witnessed the creation of the earth. Would those coming forth to earth from the city of God qualify as pilgrims?

JOB 38:4 **Where wast thou when I laid the foundations of the earth?** *declare, if thou hast understanding. ... 6 Whereupon are the foundations thereof fastened? or who laid the corner stone thereof; 7 When the morning stars sang together,* **and all the sons of God shouted for joy**?

Granted there are many questions to be answered to fill in details, but in this article we have provided a framework for the creation of the earth and the life living upon it (later articles will address more details). And that framework presents a chain of custody from *the dust of the highest part of the world* to the earth itself and to the *dust of the ground* that was used to plant Adam and other life forms on the earth. And curiously enough, this framework does not depend upon the totality of the creation process being done in twenty-four hour days approximately 6000 years ago. For we do not know the period of time that the *dust* pre-existed or the seed pattern design pre-existed before the events of Genesis 1:1 Our geologist and paleologist friends will undoubtedly have both questions and contributions to make in fitting details into that framework – that will be covered in a subsequent article in this series.

In the beginning God created the heaven and the earth. Perhaps for one to create time and space, one must exist outside of the constraints of the material world. After all, scientists are predicating that invisible 'curled up' dimensions exist—*so that things which are seen were not made of things which do appear.*

- Context Scripture Chapters*: Gen 1&2, Pro 8, Amo 5, Job 38 & 26, Rom 1, Heb 11*

Article Two

EARTH'S MAGNETIC SHIELD IN THE BIBLE?

Benefits of the "Sun and Shield"

Why do we require protection from the sun when we rely so heavily on its light and warmth for our survival? No, it's not just using a sun screen shielding to prevent getting a sun burn; it is something much more profound. We know that the NASA astronauts wore space suits when they landed on the moon — but for what purpose? Certainly, they required breathing air and also regulation of temperature within the suit for lunar temperatures varying from 260 deg. F on the 'light' side of the moon (Ooch!) to minus 280 deg. F (Brr!) on the dark side. Perhaps, the specially colored visors that they wore may give us a clue to what that 'other' needed shielding might be. Did the visors and space suit shield them from deadly radiation particles in the solar wind?

Let's fast forward to the future when some of our astronauts will land on Mars. OK, we have found water on Mars, but the atmosphere for breathing is not suitable and breathing air must be supplied. So, we will call in a fleet of Intergalactic freighters to bring in some Lox (liquid oxygen) and vaporize that for a Martian atmosphere. However, remembering the 1967 Apollo I fire resulting in the deaths of Grissom, White, and Chaffee because a pure oxygen atmosphere in the cabin ignited—perhaps we should contact Intergalactic to also get freighters of liquid nitrogen—the primary component of earth's atmosphere. This will help us get the Martian atmosphere suitable for us human beings. Like Goldilocks checking out the soup of the three bears, we don't want too much oxygen or too little oxygen, but we want it 'just right'.

After all the hard 'rehab' work and a good night's sleep plus an extra 37 minutes of 'snooze' time, we are ready to check out our revamped planet (A Mar's day is 24 hours and 37 minutes). We can live a much more leisurely life here because we have 687 days in a year — plenty of time to get things done. And, another benefit for those who hate the constant whining about us Americans being obese—we would weigh only about one-third as much on Mars (weight problem solution: Earth is too big—just move to Mars).

As we walk around, soaking up our vitamin D from the sunshine, our nerd science buddies alert us to a problem: There is a lot more cosmic radiation coming through the atmosphere than would be good for us—it could shorten our lives. Back to the launching pad—why are we having this problem on Mars and not on earth? After more researching, we are told that the magnetic field around the earth is much stronger than that on Mars. It is this 'magnetosphere' that shields the earthlings from the hard cosmic radiation particles by deflecting it past the earth. We need to solve this problem quickly, the cosmic rays are not only too harsh for us, but their 'weight' is stripping away the atmosphere that we freighted in at a cost of $780B above that budgeted by Washington. We need help—is there anyone, anywhere, who has had experience with planetary shields? We do our GigaGoogle search and come up with some very unexpected references.

No, this can't be—these references are too archaic. We need something much more modern and 'scientific' to deal with our problem. However, maybe this Mars atmospheric desolation did happen in ancient history. We need a new atmospheric 'heaven and earth' for Mars. Should we consider this archaic information after all?

> For the LORD God is **a sun and shield**: the LORD will give grace and glory: no good thing will he withhold from them that walk uprightly. O LORD of hosts, blessed is the man that trusteth in thee. PSALM 84:11

> God reigneth over the heathen: God sitteth upon the throne of his holiness. The princes of the people are gathered together, even the people of the God of Abraham: for **the shields of the earth belong unto God**: he is greatly exalted. PSALM 47:8

> For he looketh to the **ends of the earth, and seeth under the whole heaven; To make the weight for the winds**; and he weigheth the waters by measure. JOB 28:24, 25

And they that shall be of thee shall build the old waste places: *thou shalt raise up the foundations of many generations; and thou shalt be called, The repairer of the breach, The restorer of paths to dwell in.* ISAIAH 58:12

Come to think of it—there is the old saying that *If all else fails, read the directions.* Maybe we should read the manufacturer's handbook and see if we can figure out what is going on.

It does seem clear that the Bible does actually mention the relationship between the *sun* rays and the *shields of the earth.* And we know that the magnetic lines of force emanate from *the ends of the earth* and go *under the whole heaven* between the north and south poles. And scientists have found that there is a *weight for the winds* (solar winds or flares) which can strip away an atmosphere unless it is protected by the *shields of the earth*—or the shields of another planet for that manner.

The Rover pictures from Mars show a desolate planet, one that likely had rivers, seas and some form of life at one time. Did it get wasted and at some point and there will be those who *shall build the old waste places?* Maybe, someday, this question will be answered.

Now, we will go back to the mystery of why Mars does not have the same degree of magnetosphere shielding that we have on earth. Ben Weiss, professor of Planetary Science, had these comments on a National Geographic program about *Earth's Invisible Shield.*

"Unlike the earth, Mars today does not have a global magnetic field generated in its core. If you were to walk around the ancient southern highlands of Mars, you would be walking in many places in fields as strong as the earth, but they would be very complicated, almost like spaghetti pointing in all different directions, not pointing north at all.

So it is clear from that Martian crust is highly magnetized, that there must have been a magnetic field in the past, so that's the only way we know to magnetize rocks is for them to form in the presence of a magnetic field. We have some rocks that were naturally transferred from Mars to earth. These are rocks that were blasted off of Mars by an asteroid or comet impact."

Weiss went on to report on the magnetic strength of one of the most ancient of the rocks and concluded that Mars once had a magnetic field of roughly the same strength that earth has today. Fellow scientist, Jeremy Bloxham added, "When Mars lost its magnetic shield that was protecting it from the solar wind and from cosmic radiation that had consequences for the evolution of its atmosphere."

So, today we see a barren red planet with subzero temperatures, very little atmosphere and very spotty magnetic shielding. We most likely are not thankful enough that our earth lies in what is called the 'Goldilocks Zone.'

Michio Kaku, in his book, *Parallel Worlds* [ISBN 0-385-50986-3], described this zone as follows: (Chapter 8; A Designer Universe):

"WHEN I WAS A CHILD in second grade, my teacher made a casual remark that I will never forget. She said, "God so loved the earth, that He put the earth just right from the sun." As a child of six, I was shocked by the simplicity and power of this argument. If God had put Earth too far from the Sun, then the oceans would have frozen. If He had put Earth too close, then the oceans would have boiled off. To her, this meant that not only did God exist, but that He was also benevolent, so loving Earth that He put it just right from the Sun. It made a deep impact on me.

Today, scientists say that Earth lives in the "Goldilocks zone" from the Sun, just far enough so that liquid water, the "universal solvent," can exist to create the chemicals of life. If Earth were farther from the Sun, it might become like Mars, a "frozen desert," where temperatures have created a harsh, barren surface where water and even carbon dioxide are often frozen solid. Even beneath the soil of Mars one finds permafrost, a permanent layer of frozen water.

If Earth were closer to the Sun, then it might become more like the planet Venus, which is nearly identical to Earth in size and is known as the "greenhouse planet." Because Venus is so close to the Sun, and its atmosphere is made of carbon dioxide, the energy of sunlight is captured by Venus, sending temperatures soaring to 900 degrees Fahrenheit. …

To appreciate the complexity of these arguments, consider first the coincidences that make life on Earth possible. We live not just within the Goldilocks zone of the sun; we also live within a series of other Goldilocks zones. For example, our Moon is just the right size to stabilize Earth's orbit. If the Moon were such smaller, even, tiny perturbations in Earth's spin would slowly accumulate over hundreds of millions of years, causing Earth to wobble disastrously and creating drastic changes in the climate to make life impossible. …

Likewise, Earth exists within the Goldilocks zone of the Milky Way galaxy, about two-thirds of the way from the center. If the solar system were too close to the galactic center, where a black hole lurks, the radiation field would be so intense that life would be impossible. And if the solar system were too far away, there would not be enough higher elements to create the necessary elements of life."

Do we need to go back to the drawing board [manufacturer's handbook] to solve our dilemma? Is there information in the scriptures or perhaps in the ancient ruins on Mars that would give us clues? Some 'alleged discoveries' were made in the Bimini Sea on Mars of a stone like road leading to the Bimini volcano island. The stones were exactly square cut and fitted together with laser like precision. However, most scientists said that the stones had just 'evolved' into that position naturally — probably within fifty thousand years. Certainly, since marvelously complicated animal life had evolved on the earth in hundreds of thousands of years, it would be no trick for stones of this road to naturally have square faces with laser like cuts. However, scientists mulled over this question—if a Giza like pyramid were found—would it indicate natural 'evolvement' or human type building activity?

Just as we were about to give up on finding any pyramids or other type of ruins [because of the erosion and silting effect of Mar's windstorms] we hear from our intrepid archaeologist, Marianna Jones, great granddaughter of Indiana Jones. She has 'unmarred' an ancient artifact and was very excited, calling it a Rosetta Stone of creation and showing possibly how to *build the old waste places* that were ruined once Mars lost its magnetic shield. She thought it significant that it tied in with the recurring pattern of reversal of the sun's magnetic poles every 22 years – the last reversal occurring in 2012. In the same area, she also found a small fossilized **Mars** organism which she was able to fit into one of the 22 classes of **Earth** organism phyla described by James W. Valentine on page 138 in his book *On the Origin of Phyla* [ISBN-13: 978-0226845494] as follows: "The earliest major molecular phylogeny of metazoan phyla was of Field et al. (1988), who produced RNA sequences for twenty-two classes in ten phyla, quite an achievement for the state of the art at that time."

Did her twenty-two dimensions 'Rosetta Stone' provide a bridge between Astrophysics and Biophysics? But—a problem, a serious one— soon after she announced her discovery, a black Spacillac landed nearby and well-dressed men in black suits confiscated the artifact and the fossil citing 'Planational Security' as the reason for the seizure.

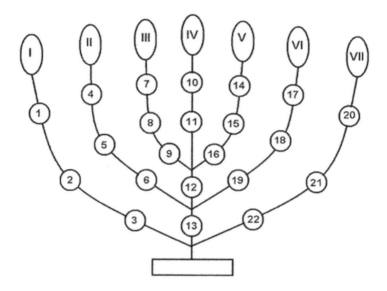

Naturally, Marianna was extremely disappointed to have the artifact taken from her, but she did remember what it looked like and immediately made the above sketch in a notebook and numbered the significant features for later reference. She surreptitiously sent the notebook sketch to her friends on earth, since she wanted to preserve the information even if she were subjected to 'memory erasure' techniques during her further debriefing by the ubiquitous agents from the black Spacillac. What did the sketch of the artifact look like?

Because of 'Planational Security' we can only reveal generic information about the above symbol—which has been described as a Rosetta stone of creation. A similar artifact found on earth with this configuration has also been 'lost' but many drawings and likenesses of it still exist. It seems this is the seven-branched candlestick with twenty-two almond bud nodes that is described in Exodus Chapter 25. The seven branches are said to represent the Seven Spirits of God outlined in Isaiah Chapter 11 and the twenty-two almond buds represent the twenty-two letters of the Hebrew alphabet as fully outlined in Psalm 119. Apparently, these twenty-two letters were used as the 'language of creation' and involved in words such as *"Let there be Light.* So, knowledge of these would be necessary if we wanted to *build the old waste places* that now exist on Mars.

For those having further curiosity about the candlestick, check out the book, *The Dove Code* [IBSN 978-158-169-2990], by Chatan N. Kallah, and examine information in Chapter 4 and also in the Appendices. We will list here a brief quote from the book about the 'hidden dimensions' symbolized in the candlestick. Sarah is telling her friend Kim about the Hebrew alphabet:

""Right in the middle of the Bible is Psalm 119. It is an acrostic, and each of the twenty-two letters of the Hebrew alphabet has eight verses that begin with that letter," said Sarah. "Those little words that you see are the twenty-two letters of the Hebrew alphabet."

Kim's jaw dropped and she gasped, "I can't believe it! Not only do you have the seven curled up, invisible, dimensions in the candlestick, but you also have the twenty-two Hebrew alphabet letters. These must be the building blocks of creation! **The candlestick could truly be the Rosetta stone between the mysterious eleven (7 + 4) dimensions and the twenty-six (22 + 4) dimensions that quantum physicists are feverishly trying to decipher.** The four dimensions that we can see—you know—are length, width, height, and time. The rest are invisible to our eye. Amazing! Absolutely Amazing!"

Could this be the information that we need to build a habitable and safe atmosphere on Mars? Perhaps it is, but for the time being, unless we are given additional clearance to publish other than generic details, we will keep this information to ourselves—and pass off this writing as just a 'made up' little story—kind of like the one about Goldilocks and the Three Bears.

- Context Scripture Chapters: *Psa 84 & 47, Job 28, Isa 58, Gen 1, Psa 119*

Article Three

ARE YOU SMARTER THAN A PhD THEORETICAL PHYSICIST?

Maybe Not—Unless You Have Secret Notes

If you are a PhD theoretical physicist reading this, you might find the question somewhat insulting. However, if you simply have a curiosity about the universe, then the question is more in the intriguing category. So, let's look at some of the questions about the universe that our PhD's are struggling with and see if we, with a little help, can help solve the problems.

But, you may say, while I love science, neither science nor math is my strong suit. How could I solve puzzles about the universe? Suppose that you are blindfolded, taken in a car to a strange place, and led into a room. Without taking off the blindfold, you are asked to describe the contents of that room. Well, if it's a kitchen you might smell or taste food. If it's a living room, you might hear a TV. If it's a bedroom, you might feel the soft texture of a pillow. But, whatever kind of room it is, you would say to yourself, "Oh, if I could just look around for a minute or so, I would be able to describe it so much better!" Or, maybe it's like the childhood game where we try to find a hidden object — if someone would only tell us whether we are 'warm' or 'cold' we would find it much sooner.

So, let's take a look at a problem that scientists have been feverishly working on – it involves string theory and M-theory (which some call membrane theory). Let's try to keep a very complex subject as simple as we can. Scientists are beginning to think that the whole plethora of different sub-atomic particles (such as quarks, anti-quarks, squarks, bosons, gluons, photons, et cetera, ... et cetera ... et cetera ... et cetera) are not different particles at all, but are like different notes played on a violin string or a string instrument of some type. Then, perhaps a membrane is like a piano with all types of different vibrating strings and configurations attached to it. Think of all the tunes you can play on the piano and this gives one a concept of why particle physics is so complicated. And maybe a 'multi-brane' is like a whole symphony orchestra of galaxies and / or universes. Granted, these examples are way too simplistic and would likely be criticized by the PhDs, but we have to start somewhere.

Scientists around the world, at places like Cern in Switzerland and Fermilab in Chicago are expending great energies in hot pursuit of exploring these theories. In addition, the mathematicians are poring over their formulas trying to probe the frontiers of string and M-theory. At this point, perhaps the mathematicians can go where no particle collider (like the one at Cern) can yet go—due to energy limitations. Einstein was fond of pointing out to his friends that without expensive laboratories or equipment, he could do his experiments mathematically in his mind.

So, let's examine a problem the PhD's are working on—and perhaps—in our simplistic way—we might be of assistance. Maybe not — but at least it would be fun to try. It involves the confusion that scientists are having trying to decide whether there are eleven dimensions or twenty-six dimensions. Don't panic—we will try to give a simplistic description of a 'dimension' in just a bit. The truth is stranger than fiction—so lest you think we are making this stuff up, we will do a search on 'eleven dimensions and twenty-six dimensions' to find which sources mention both the eleven and twenty-six. Here are just a few of the 'fish' that we have scooped up in our Google net:

> **the nth** <u>dimension</u> » **tomorrow's theories** » **types of string theories**
> **Counterclockwise vibrational patters occupy** twenty-six dimensions, ... **M-theory demands** eleven dimensions (**ten space** dimensions **and one time** dimension) ...
> *library.thinkquest.org/04apr/01330/.../typesof_st.htm*
> - Cached - Similar
>
> **Kaluza–Klein unification – Some possible extensions - Elsevier**
> In 11 dimensions **Kaluza–Klein represents more or less a super symmetric gravity theory,** .... [4] **Ji-Huan He,** Twenty-six **dimensional polytops and high energy** ...
> *linkinghub.elsevier.com/retrieve/pii/*

S0960077907008272 - Similar
by MS El Naschie - 2007 - Cited by 3 - Related articles

Compactification (physics) - Wikipedia, the free encyclopedia
8 Jul 2009 ... At the limit where the size of the compact **dimension** goes to zero, ... **eleven**, or **twenty-six dimensions** theoretical equations lead to ...
en.wikipedia.org/wiki/Compactification_(physics) - Cached - Similar

String Theory Dimensions **Isbn Theories Extra Superstring**
... **twenty-six**. More precisely, the bosonic string theories are 26-dimensional, while superstring and M-theories turn out to involve 10 or 11 **dimensions**. ...
www.economicexpert.com/a/String:theory.htm - Cached - Similar

OK, in order to get the 'flavor' of this problem, we have netted five references and could add many more if we worked at it. But, what is the nature of eleven dimensions and/or twenty-six dimensions and why do scientists seem to vacillate back and forth between 26 dimension string theory and 11 dimensions M-theory?

First, we will define what we mean by dimensions. Let's suppose that you are wearing a wristwatch with a rectangular face. Let's say the face is one half inch wide and we then let x = 0.5 inch. We measure the height of the watch face as three-fourth inch high and then let y = 0.75 inch. We measure the thickness of the watch and find it is one eighth inch thick and then let z = 0.125 inch. We have measured the three dimensions of width, height, and thickness and named these dimensions x, y, and z. What could the fourth dimension be? We look at our watch and note that the time is 11:30 AM and then let t = 11.5 hours. Now, we have defined x, y, z and t, a total of four dimensions that we are all familiar with and we 'see' each and every day. These are our readily observed dimensions.

Where do the eleven dimensions and the twenty-six dimensions come from? Well, scientists think that eleven dimensions exist in M-theory—four that we can observe—and seven that are hidden. The hidden dimensions are 'curled up' so we can't see them — sort of like seven 'ghosts' that are mathematically and intrinsically there, but we just can't see them with our natural eyes. However, to keep this from getting too ethereal, we might point out that the so-called fifth dimension is thought to be electromagnetism. Remember your science class experiment with two magnets? Two magnets were put under a piece of paper and iron filings were sprinkled on the paper. You couldn't see the magnetic lines of force between them but you could see their effect in aligning the iron filings along their force lines.

In reality, or unreality—which ever you prefer—the hidden dimensions are simply those dimensions that we don't see with our natural eyes. So how many hidden dimensions are there? For M-theory, a number of scientists think there are 11 total dimensions minus four seen dimensions. For string theory, there are 26 total dimensions minus four seen dimensions. In summary **11 – 4 = 7** and **26 – 4 = 22**.

But this is getting way too confusing—one feels like they are in a dark room groping for knowledge. Oh, if someone would only light a candle so that we could see into one of these mysterious, hidden dimensions. Or, better yet, light seven candles, so we could see all seven hidden dimensions – or maybe 22 hidden dimensions — whichever or whatever it is. So, let there be light!

What is this thing? Where did it come from and what does it represent?

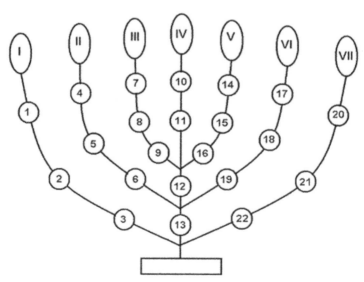

Well, OK, we did ask for seven candles to light up the seven hidden dimensions and we do see candles I through VII. And we were interested in 22 nodes for our vibrating strings and we see that this symbol, or whatever it is, has 22 nodes in it. Is this a bridge between M-theory and string theory? Is this something that many of the PhDs have overlooked and yet we can 'see' the connections in this symbol?

Before we go further, let's look at some quotes from the writing of an excellent author who is in the frontiers of string theory and M-theory. Brian Greene is a professor of physics and mathematics at Columbia University. He wrote *The Elegant Universe* and also *The Fabric of the Cosmos* [IBSN: 0-375-72720-5]. In Chapter eleven of the second book, he writes about Edward Witten, a pioneer in string and M-theory:

"Eleven Dimensions

So, with our newfound power to analyze string theory, what insights have emerged? There have been many. I will focus on those that have had the greatest impact on the story of space and time.

Of primary importance, Witten's work revealed that the approximate string theory equations used in the 1970s and 1980s to conclude that universe must have nine space dimensions missed the true number by one. The exact answer, his analysis showed, is that the universe according to M-theory has ten space dimensions, that is, eleven space-time dimensions. Much as Kaluza found that a universe with five space-time dimensions proved a framework for unifying electromagnetism and gravity, and much as string theorists found that a universe with ten spacetime dimensions provided a framework for unifying quantum mechanics and general relativity, Witten found that a universe with eleven space-time dimensions provided a framework for unifying, all string theories.

... While Witten's discovery surely fit the historical pattern of achieving unity through more dimensions, when he announced the result at the annual international string theory conference in 1995, it shook the foundations of the field."

We can see an emerging pattern of scientists over the last century edging ever closer to the seven hidden dimensions and twenty-two hidden dimensions that we have shown in our symbolic drawing. But, what is this thing? Does it now exist? We don't know – it may be lost. However, some of you will instantly recognize it from another field since many replicas and written descriptions of it still exist today.

There is a story that has been around for some years about scientists eagerly climbing the mountain of knowledge and breathlessly anticipating their first look to see what lies on the other side of the mountain that has been blocked from their view. As they reach the top of the mountain and look over the '**other side**', they see a group of theologians who ask them, "What took you so long? We have been here waiting for you."

We will now go to an ancient text to see if we can find a description of this thing that might symbolize the seven hidden dimensions of M-theory and the twenty-two hidden dimensions of string theory. We quote from the archaic text (high lights added):

*And thou shalt make a **candlestick of pure gold**: of beaten work shall the candlestick be made: his shaft, and his branches, his bowls, his knops, and his flowers, shall be of the same. And six branches shall come out of the sides of it; three branches of the candlestick out of the one side, and three branches of the candlestick out of the other side: **Three bowls made like unto almonds, with a knop and a flower in one branch; and three bowls made like almonds in the other branch, with a knop and a flower: so in the six branches that come out of the candlestick. And in the candlestick shall be four bowls made like unto almonds,** with their **knops** and their flowers. And there shall be a knop under two branches of the same, and a knop under two branches of the same, and a knop under two branches of the same, according to the six branches that proceed out of the candlestick. Their knops and their branches shall be of the same: all it shall be one beaten work of pure gold. And thou shalt make **the seven lamps** thereof: and they shall light the lamps thereof, that they may give light over against it.*

The Articles of Configuration

The Hebrew Candlestick in the Tabernacle in the Wilderness
Courtesy of Gerth Median Gmbh; Postfach 1148; 3567; Asslar, Germany
(Note: *The Tabernacle of God in the Wilderness* is currently out of print)

Hey, you might say—this sounds like something from the Bible, and that it is — from Exodus Chapter 25. Now let's go through the math. There are three bowls made like almond buds on each of six branches. So, 6 X 3 = 18. Then, there are four buds on the central shaft of the candlestick. So, 18 + 4 = 22 buds or perhaps we would call it nodes. Then, there as seven lamps, what could the lamps represent? Again, we will resort to an ancient text as written by the Apostle John in the book of Revelation: *And out of the throne proceeded lightnings and thunderings and voices: and there were* **seven lamps of fire** *burning before the throne,* **which are the seven**

Spirits of God. (Revelation 4:5) Putting it all together, we have 22 nodes and seven lamps. So what, what does that prove?

Loggers will tell you the strategy in relieving a log jam is to find the key log, and once that log is found and jarred loose—the whole raft of logs once again begins to float down the river. So, what is our key log — why is the seven lamped and twenty-two noded candlestick significant? Since we seem to be switching gears into the theological arena, we ask how the world came into being as a way of finding our key log. We have the following excerpts from Paul, John, and Moses:

*Through faith we understand that the worlds were framed by the word of God, so **that things which are seen were not made of things which do appear**. In the **beginning was the Word**, and the Word was with God, and the Word was God. The same was in the beginning with God. All things were made by him; and without him was not any thing made that was made. In the beginning God created the heaven and the earth. And the earth was without form, and void; and darkness was upon the face of the deep. And the **Spirit of God** moved upon the face of the waters. And God said, **Let there be light: and there was light**.*

So, if the four dimensions (*things which are seen*) are made by seven hidden dimensions (*not of things which do appear*), and the words (*Let there be light*) were spoken, what is necessary to form words? One would need an alphabet to form words. Since the story of creation is written in Hebrew, it would seem likely that the words of creation were spoken in Hebrew. How many letters would you guess are in the Hebrew alphabet? Strangely enough, the Hebrew alphabet is given in the longest chapter in the Bible — Psalm 119. These are given in an acrostic — the first eight verses begin with the letter Aleph, the second eight begin with the letter Beth — and so on through the alphabet until we reach the 22nd letter Tau. Have we broken loose the key log in the logjam so that revelation of some of the very secrets of both science and theology will come forth? For the scientists tell us that there are 22 hidden dimensions that show up in string theory and there are Seven Spirits within the envelop of the Spirit of God which moved upon the waters. Maybe we are on to something. Is there some kind of code in all this?

In the book, *The Dove Code* [ISBN 978-158169-2990] by Chatan N. Kallah, there is a dialog where Sarah is explaining the candlestick puzzle to her scientifically oriented college roommate, Kim. In Chapter 4, we read the following:

"Oh, those," replied Sarah. "You see the candlestick has twenty-two almond flowers engraved on it. Six branches have three flowers apiece, which gives eighteen, and then there are four on the main shaft. That gives a total of twenty-two. The almond rod was what Moses used to part the Red Sea. You know what the twenty-two almond flowers represent, don't you?"

"Not a clue—tell me."

"Right in the middle of the Bible is Psalm 119. It is an acrostic, and each of the twenty-two letters of the Hebrew alphabet has eight verses that begin with that letter," said Sarah. "Those little words that you see are the twenty-two letters of the Hebrew alphabet."

Kim's jaw dropped and she gasped, "I can't believe it! Not only do you have the seven curled up, invisible, dimensions in the candlestick, but you also have the twenty-two Hebrew alphabet letters. These must be the building blocks of creation! **The candlestick could truly be the Rosetta stone between the mysterious eleven (7 + 4) dimensions and the twenty-six (22 + 4) dimensions that quantum physicists are feverishly trying to decipher.** The four dimensions that we can see—you know—are length, width, height, and time. The rest are invisible to our eye. Amazing! Absolutely Amazing!"

"Oh, there's a lot more," said Sarah. "I'll show you some of Uncle Andy's and Aunt Myra's emails on this subject. They discuss how the things we see were framed by faith from the unseen things of the spirit.

Let's see—let me do a quick search on one of my favorite sites that I use when you snow me with all your technical jargon. Here: http://www.explainthatstuff.com/ searching under voice recognition. Ah, here is the pertinent part:"

Voice recognition programs can recognize people's speech through a combination of these techniques. Just as written words in English are made up of 26 possible letters, so spoken words are made up of 44 possible sounds known as **phonemes**. Crudely speaking, phonemes correspond to the syllables in words (there's a bit more to phonemes than that, but that's an easy way to think of it). In theory, a computer could understand anything you said if you trained it to recognize the 44 basic phonemes. Put another way, if you spoke any word, all your computer would have to do would be to split the overall word sound into its component phonemes, identify what letter sounds those phonemes represented, and then it would be able to figure out the word. voicerecognition.html

"Perhaps when you combine the twenty-two phoneme sounds of the patrix with the twenty-two phoneme sounds of the matrix, you get the spoken Word which in the beginning framed the worlds."

The Dove Code goes on to list other 'revelations' – particularly those given in the book's Chapter 15 on 'membrane' stretching (*O LORD my God, thou art very great; thou art clothed with honour and majesty. Who coverest thyself with light as with a garment: who stretchest out the heavens like a curtain:*) and reguarding DNA in the book's appendix (see http://TheAncientChest.com/)

So, we have come full circle back to our original question: Are you smarter than a PhD theoretical physicist? And most of us would have to say probably not—unless we were given revelations from our manufacturer's handbook. After all, *when all else fails, read the directions* and it would seem that our manufacturer knows more about what makes us tick than we do.

However, scientists need theologians, and theologians need scientists so that we may all progress together—otherwise we have elitism. The Apostle Paul wrote: *If the foot shall say, Because I am not the hand, I am not of the body; is it therefore not of the body? And if the ear shall say, Because I am not the eye, I am not of the body; is it therefore not of the body? If the whole body were an eye, where were the hearing? If the whole were hearing, where were the smelling? But now hath God set the members every one of them in the body, as it hath pleased him. And if they were all one member, where were the body? But now are they many members, yet but one body. And the eye cannot say unto the hand, I have no need of thee: nor again the head to the feet, I have no need of you.* ***Nay, much more those members of the body, which seem to be more feeble, are necessary:*** Is not the little child who can 'see with the eye of faith' just as important as the scientist with a sophisticated measuring machine?

- Context Scripture Chapters: *Exo 25, Heb 11 , Rev 4, Psa 104, ICor 12*

Article Four

THE LIGHT AT THE TOP OF THE MOUNTAIN

A Chronology of the Path to the Unseen Dimensions

By definition physicists are most interested in the *physical* world. After all, *physicists* who study *physics* want everything to be proven logically and shy away from 'myths'. Things need to be *seen* and reasoned out. However, a strange thing happened to scientists on their way to the forum of validation of ideas. Einstein, famous for his theory of relativity, spent the later part of his life searching for the *Holy Grail* of science—a theory of everything. Einstein had jarred scientists out of their comfort zone of the physics of Isaac Newton which only basically described the four dimensions of length, width, depth and time. After Einstein, string theory came into vogue in the 1970's. It was thought that 'particles' consisted of super tiny vibrating violin like strings which manifested different characteristics depending on how the string was stimulated.

For a while it was thought that string theory was TOE, an acronym for the **T**heory **O**f **E**verything. String theorists were proposing that there were unseen dimensions that were somehow 'curled up' in such a way that we couldn't see them. A common example given was that of the shadow of a garden hose. The shadow would appear flat on a surface—such that if we only saw its shadow, we would think the hose was flat rather than being curled up into a cylinder. It seems that in their search for TOE's *Holy Grail*, scientists had inadvertently run into the scripture describing the symbols of the furnishings of the tabernacle *Who serve unto the example and shadow of heavenly things, as Moses was admonished of God when he was about to make the tabernacle: for, See, saith he, that thou make all things according to the pattern shewed to thee in the mount.*

But another funny thing happened to the scientists on their way to the forum of the validation of ideas. The mathematics of string theory predicted that in addition to the four *seen* dimensions of length, width, depth, and time, there were six *unseen* dimensions—giving a total of ten seen and unseen dimensions. For a while, this was fine in that TOE seemed within grasp. Then, it seemed things began falling apart for string theory—so many variations of the theory began popping out of the woodwork that scientists began to lose faith in an overall string theory. As one professor described all the variant theories, it was 'an embarrassment of riches'. Now the stage is set for our narrative describing the tenth dimension's search for the eleventh dimension.

Is it possible that the unseen dimensions are spiritual dimensions? For example, scientists have identified one of the unseen dimensions as the electromagnetic dimension. While we don't see magnetic lines of force, we can see their effect in arranging iron filings around a magnet. We can see the effect of lightning arcing through the air. One of the clues from antiquity that electromagnetism is a spiritual dimension was penned by the Hebrew prophet Job: *In thoughts from the visions of the night, when deep sleep falleth on men, Fear came upon me, and trembling, which made all my bones to shake. Then* **a spirit passed before my face; the hair of my flesh stood up:** Scientists may debate what a spirit is (or even if a spirit exists at all) but it is no secret that situations arise where our hair stands up on end due to some electromagnetic force. And it is no secret, that when we are truly inspired (*in spirited*), we may experience 'goose bumps'. It does appear that a spirit is still a spirit whether its intentions are good or evil.

In about 2000, scientists were becoming increasingly disillusioned with string theory due to the multiplicity of theories of everything that it presented. But it appears that not everybody was saddened by this—because there were those who were promoting what was thought to be a competing theory—namely super gravity. Super gravity required eleven dimensions while string theory proposed ten dimensions. It was thought that force of gravity was distributed through the other dimensions in such a way that its force was somewhat diluted in the earth realm. So, the eleven dimension super gravity scientists were at war with the ten dimension string theorists.

Imagine that you are standing upon the top of a high mountain in the Andes looking eastward across the vast jungle of the Amazon basin. Toward the northeast you see movement in the jungle—it looks like a large crew is blazing a trail toward the mountain. And then off to the southeast you see a smaller crew also blazing a trail toward the mountain. Year by year they come closer to the mountain—you think that it is a shame that these pioneers are separated from each other. If they could but get together, they would make much faster progress. Using this allegory, we are ready to present a chronology of the progress of scientists working their way to the light at the top of the mountain. The 'logbook' entries of the two trails are as presented here.

The following is a transcript of the dialog from a portion of an excellent BBC / Horizon production *Parallel Universes* which was shown on the Science Channel. It also can be viewed at: http://www.youtube.com/watch?v=Z7SDrj4Tjvk

The speakers are the narrator and also professors Michael Duff (University of Michigan; Imperial College London), Burt Ovrut (University of Pennsylvania), Michio Kaku (City College of NY) and Paul Steinhardt (Princeton University).

****

NARRATOR: When string theory fell apart, not everyone was distraught – some people seemed to even relish the fact.

DUFF: If string theory was really this so-called theory of everything, five theories of everything seemed like an embarrassment of riches.

NARRATOR: Michael Duff had been the rising star of an earlier idea called super gravity. String theory had displaced it and almost destroyed Duff's career.

DUFF: Physics tends to be dictated by fad and fashion. There are the gurus who dictate the direction in which new ideas grow. It was a very lonely time in many ways. I tried to get graduate students interested, many of them would say, "Well, look, you may be right and you may be wrong, but if I work in super gravity, I may not find a job.

NARRATOR: What made the experience of the super gravity guys so galling was their theory wasn't so very different from string theory to begin with? In fact the main point of disagreement between them was a point of detail which could seem like nitpicking. It was about the number of dimensions in the universe. We normally think of ourselves as living in a three dimensional world—right or left, up or down, and forward and backward. But physics liked having extra dimensions. Einstein suggested time should be a fourth dimension. Then someone suggested a fifth spatial dimension and then a sixth. The numbers just kept growing. The extra dimensions were spaces in the universe which we could never perceive.

Most were microscopically small but scientists believed they were really there. String theory had been convinced that there were exactly ten dimensions.

OVRUT: Now if you have a little oscillating string, it has to have enough room to oscillate properly. And when one works this out mathematically you find it – it has got a very clear answer—it just had to be in ten dimensional spaces.

KAKU: Ten Dimensions!

OVRUT: Nine spatial dimensions and one time!

NARRATOR: Super gravity though had been convinced that there were exactly eleven dimensions.

DUFF: The equations of super gravity took the simplest and most elegant form when written in this eleven dimensional framework.

KAKU: There was a war between the tenth dimension and the eleventh dimension. In the ten dimension bandwagon we had string theorists – hundreds of them working to tease out all the properties of the known universe from one framework — a vibrating string. And then we have this small band of outlaws working in the eleventh dimension.

NARRATOR: While string theory was in its ascendancy, few took seriously the eleventh dimension, but the super gravity guys never gave up hope.

DUFF: I did—at the bottom—always feel convinced eventually eleven dimensions would have its day. I wasn't sure when, I wasn't sure how, but I was always convinced sooner or later eleven dimensions would be seen to be at the heart of things.

NARRATOR: But by now, the boot was on the other foot. String theory was in trouble. Its five different versions meant it couldn't be the all-embracing theory physics was looking for. Everything it seems had been tried to save string theory. Well, almost everything.

DUFF: An astonishing announcement was made!

KAKU: There was yet another shockwave that revolutionized the whole landscape!

NARRATOR: In a final desperate move, the string theorists tried adding one last thing to their cherished idea. They added the very thing they had spent a decade rubbishing: the eleventh dimension. Now something almost magical happened to the five competing theories.

OVRUT: The answer turned out to be remarkable – and it was really absolutely remarkable, it turns out they were the same! These five string theories turned out to be simply different manifestations of a more fundamental theory.

KAKU: In eleven dimensions — looking from the mountain top — looking down you could see string theory as being part of a much larger reality — the reality of the eleventh dimension.

DUFF: Well, it was a wonderful feeling to feel that those years spent in the eleventh dimension were not completely wasted.

NARRATOR: The two camps had been absolutely certain the other was wrong. Now, suddenly, they realized their ideas complimented each other perfectly. With the addition of one extra dimension, string theory made sense again. But, it had become a very different kind of theory.

OVRUT: What happed to the string?

NARRATOR: The tiny invisible strings of string theory were supposed to be the fundamental building blocks of the matter in the universe. But now, with the addition of the eleventh dimension, they changed. They stretched and combined. The astonishing conclusion as that all the mater in the universe was connected to one membrane. In effect, our entire universe is a membrane. The quest to explain everything in the universe would begin again and its heart would be this new theory. It was dubbed M-theory or Membrane theory.

****

As one reads the logbooks of the two journeys toward the theory of everything, the words spoken to Moses on the mount ring out across the land: *See, saith he, that thou make all things according to the pattern shewed to thee in the mount.* Upon the mountain top, Moses was shown the pattern of furnishings for the tabernacle to be built in the wilderness. One of the furnishings in this tabernacle was the seven lamped candlestick — a diagrammatic sketch of this is shown in the following illustration. Also, shown are the twenty-two almond blossoms which represent the twenty-two letters of the Hebrew alphabet.

The Light At The Top Of The Mountain

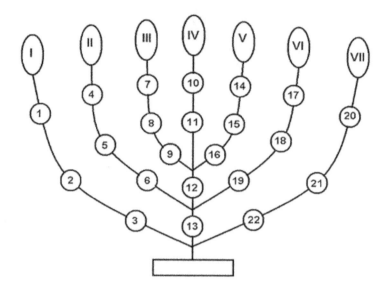

Faith: *Through faith we understand that the worlds were framed by the word of God, so that **things which are seen were not made of things which do appear.***

The Instructions to Moses: *And thou shalt make a candlestick of pure gold: of beaten work shall the candlestick be made: his shaft, and his branches, his bowls, his knops, and his flowers, shall be of the same. **And six branches shall come out of the sides of it;** three branches of the candlestick out of the one side, and three branches of the candlestick out of the other side:*

Seven Spirits of God: *And out of the throne proceeded lightnings and thunderings and voices: and there were **seven lamps of fire** burning before the throne, which are the **seven Spirits of God**.*

The words of Professor Michio Kaku had a prophetic ring when he described the meeting of string theory and super gravity in the following way: "In eleven dimensions — looking from the mountain top — looking down you could see string theory as being part of a much larger reality— the reality of the eleventh dimension." One is reminded of an old story about scientists eagerly climbing the mountain of discovery to see what was one on the other side. As they peaked over the top they saw a group of theologians that had been patiently waiting for them.

The candlestick has one central shaft and six branches. The string theorists were like six branches (five spatial dimensions plus one time dimension) looking for a common tie-in to a central trunk. The theory of the six branches fell apart because it was missing the main shaft of the candle stick. The super gravity group had the main trunk of the candlestick but it was a very lonely existence without branches for the trunk – sort of a lonely garden of Gethsemane. And then—finally!—the branches found the central trunk and all fell into place!

I am the vine, ye are the branches: He that abideth in me, and I in him, the same bringeth forth much fruit: for without me ye can do nothing.

We are reminded of the Psalms where King David writes music for the ten strings of the harp. Without the main frame of the harp, the ten strings simply do not make music. *Praise the LORD with harp: sing unto him with the psaltery and an instrument of ten strings.* But, this is surely just a co-incidence—the scriptures would not use a symbol of one thing to represent some underlying truth. Or, would they? David wrote: *I will incline mine ear to a parable: I will open my dark saying upon the harp.*

Well, what about the view from the mountaintop? We rejoin the dialogue of *Parallel Universes* for a description.

****

NARRATOR: With M theory, we see that at last there was a theory which might explain everything in the universe. But, before they could decide if this was true, the scientists needed to learn more about this new eleventh dimension. It quickly became clear that it was a place where all the rules of common-sense have been abandoned. For one thing it is both infinitely long but only a small distance across.

STEINHARDT: That eleventh dimension will, at its maximum size, could be something like a trillionth of a millimeter.

OVRUT: Well, this is ten to the minus 20th of a millimeter. That's taking a millimeter and dividing it by ten with twenty zeroes after it —that's very, very small.

NARRATOR: That means it exists only one trillionth of a millimeter from every point in our three dimensional world. It's closer than your clothes to your body and yet we can't sense it. In this mysterious space, our membrane universe is floating. At first no one could imagine how it worked.

****

No doubt the above statements had everyone scratching their head and wondering, 'How can this be?" How can an UNKNOWN world not be seen and yet be so close to us. Again we resort to ancient scripture to see if this riddle has been described before.

For as I passed by, and beheld your devotions, I found an altar with this inscription, TO THE UNKNOWN GOD. Whom therefore ye ignorantly worship, him declare I unto you. ... Neither is worshipped with men's hands, as though he needed any thing, seeing he giveth to all life, and breath, and all things; And hath made of one blood all nations of men for to dwell on all the face of the earth, and hath determined the times before appointed, and the bounds of their habitation; **That they should seek the Lord, if haply they might feel after him, and find him, though he be not far from every one of us:** *For in him we live, and move, and have our being; as certain also of your own poets have said, For we are also his offspring.*

Could it be that *not far from every one of us* could be one trillionth of a millimeter?

The *Parallel Universes* dialog goes on to describe the possibility that our universe may be paralleled by others. (Hopefully, our readers will make time to watch this show in its entirety). Does this mean that there can be other realms besides the earth realm? The scriptures speak of at least three heavens and two hells.

I knew a man in Christ above fourteen years ago, (whether in the body, I cannot tell; or whether out of the body, I cannot tell: God knoweth;) such an one caught up to **the third heaven.**

For great is thy mercy toward me: and thou hast delivered my soul from **the lowest hell.**

You have heard the phrases 'hell on earth' or 'heaven on earth' used to describe the low and high experiences that mankind goes through. *Parallel Universes* describes a possibility that is almost too mind boggling for both scientists and theologians to think about — that of parallel existence of each of us in different universes. Stock up on a good supply of Excedrin before doing serious thinking about that possibility.

And then, what about the idea of the universe consisting of tiny strings attached to a stretchable membrane? Again, it is written: *It is he that sitteth upon the circle of the earth, and the inhabitants thereof are as grasshoppers;* **that stretcheth out the heavens as a curtain***, and spreadeth them out as a tent to dwell in:*

But, stretching out an expanding universe is a whole new subject and the scriptures have much to say about this. This will be discussed in another article.

- Context Scripture Chapters: *Heb 8&11, Job 4, Exo25, Rev 4, Joh 15, Psa 33 & 49, Act 17, IICor 12, Psa 86, Isa 40*

Article Five

EXPANDING UNIVERSE IN SCRIPTURES?
Or—Is It In A Steady State?

The Opinions of Einstein, Hoyle and God

A long time ago in a place far away our universe began by expanding outward from a common origin—so say most scientists today. However, this consensus has been reached more recently as a result of improved tools for observing the universe.

Strangely enough, two towering figures of science, Albert Einstein and Fred Hoyle clung tenaciously to the idea that the universe always existed in a steady state. Albert Einstein's contributions to the concepts of relativity are so well known that it would only belabor the point to try to list them here. Fred Hoyle, although not as well-known as Einstein, made major contributions in helping us to understand the role of the stars in the forming the nucleus of the very chemical elements that make up our universe. What is the opinion of God? It would be highly, highly presumptuous for any one of us to say that we have the totality of the mind of God on this, but it would be instructive to review writings given in the Bible to find clues. Since the Bible begins with the words, *In the beginning* … it would seem reasonable to say that while God may have always existed, God would tell us that this universe had a beginning. But setting aside the question of the initial creation—once that happened—is the universe in a steady and relatively unchanging state? We will delve into opinions that were advanced by Einstein and Hoyle and examine some surprising – and almost shocking — statements from the Bible that were written long before the time of Galileo, Newton, Einstein and Hoyle.

Sometimes, scientists humorously refer to the UFF which is short for the Universal Fudge Factor—defined as: The factor by which you manipulate your data to get the right answer. Curiously, many of the breakthroughs are made when someone takes the time to find out why the UFF works. It seems that Einstein started out with a belief in a steady state universe otherwise known as a static universe, but to prove his point, he had to resort to utilizing a UFF. Well-known astrophysicist Stephen Hawking recounts the story in Chapter 7 of his book *A Briefer History of Time* [ISBN –13:978-0-553-80436-2].

"Yet so strong was the belief in a static universe that it persisted into the early twentieth century. Even Einstein, when he formulated the general theory of relativity in 1915, was so sure that the universe had to be static that he modified his theory to make this possible by introducing a fudge factor, called the cosmological constant, into his equations. The cosmological constant had the effect of a new "antigravity" force, which, unlike other forces, did not come from any particular source but was built into the very fabric of space-time. As a result of this new force, space-time had an inbuilt tendency to expand. By adjusting the cosmological constant, Einstein could adjust the strength of this tendency. He found he could adjust it to exactly balance the mutual attraction of all the matter in the universe, so a static universe would result. He later disavowed the cosmological constant, calling this fudge factor his "greatest mistake." As we'll soon see, today we have reason to believe that he might have been right to introduce it after all."

Hawking goes on to explain that physicists have postulated that 'dark energy' may be at the root of the paradox as to why the universe is expanding:

"And it is very strange, since the effect of the matter in space, whether high or low density, can only be to slow the expansion. Gravity is, after all, attractive. For the cosmic expansion to be accelerating is something like the blast from a bomb gaining power rather than dissipating after the explosion. What force could be responsible for pushing the cosmos apart ever faster? No one is sure yet, but it could be evidence that Einstein was right about the need for the cosmological constant (and its anti-gravity effects) after all."

Along this line, David the Psalmist wrote: *Yea, the darkness hideth not from thee; but the night shineth as the day: the darkness and the light are both alike to thee.* Also, Moses wrote: *And God saw the light, that it was good: and God divided the light from the darkness.*

So, now we will take a look at the steady state theory 'according to Hoyle.' Hoyle graduated from Cambridge University. Hoyle's fame soared when he gave science lectures on BBC and engaged in debates with

George Gamow, a Russian scientist and a champion of the non-steady big bang theory. Of course, the big bang theory was diametrically opposite to Hoyle's steady state theories. Later, in 1965, when residual microwave background radiation from the 'big bang' was discovered, Hoyle's steady state theory fell out of favor. Michio Kaku, in Chapter 3 of his book *Parallel Worlds* [IBSN 0-385-50986-3], gives this account:

"Hoyle never shied away from a good fight. In 1949, both Hoyle and Gamow were invited by the British Broadcasting Corporation to debate the origin of the universe. During the broadcast, Hoyle made history when he took a swipe at the rival theory. He said fatefully, "These theories were based on the hypothesis that all matter in the universe was created in one big bang at a particular time in the remote past." The name stuck. The rival theory was now officially christened "the big bang" by its greatest enemy. ...

The discovery of the microwave background by Penzias and Wilson had a decided effect on the careers of Gamow and Hoyle. To Hoyle, the work of Penzias and Wilson was a near-death experience. Finally in Nature magazine in 1965, Hoyle officially conceded defeat, citing the microwave background and helium abundance as reasons for abandoning his steady state theory. But what really disturbed him was that the steady state theory had lost its predictive power. "

To Hoyle's credit and also to Gamow's credit, both were well recognized in developing the theories of how the elements were synthesized in the cores of stars. Gamow displayed a humorous and poetic streak when he wrote:

> *There was a young fellow from Trinity*
> *Who took the square root of infinity*
> *But the number of digits*
> *Gave him the fidgets;*
> *He dropped Math and took up Divinity*

Now that we have discussed some of the opinions of Einstein and Hoyle, what about the opinions of God? Do you think that if the astrophysicists took up divinity as a co-major, 'discoveries' might come just a bit easier? Yes, we have said that God's opinion would be the universe has an *"In the beginning ..."*. However, what other clues can we find in the scriptures? Before we go there, let us examine excerpts from other chapters in Kaku's book *Parallel Worlds*:

" "SPECTACULAR REALIZATION,"
Alan Guth wrote in his diary in 1979. He felt exhilarated, realizing that he might have stumbled across one of the great ideas of cosmology. Guth had made the first major revision of the big bang theory in fifty years by making a seminal observation: he could solve some of the deepest riddles of cosmology if he assumed that the universe underwent a turbocharged hyperinflation at the instant of its birth, astronomically faster than the one believed by most physicists. With this hyperexpansion, he found he could effortlessly solve a host of deep cosmological questions that had defied explanation. It was an idea that would come to revolutionize cosmology. (Recent cosmological data, including the results of the WMAP satellite, are consistent with its predictions.) It is not the only cosmological theory, but is by far the simplest and most credible. ...

In the inflationary scenario, in the first trillionth of a trillionth of a second, a mysterious antigravity force causes the universe to expand much faster than originally thought. The inflationary period was unimaginably explosive, with the universe expanding much faster than the speed of light. ... to visualize the power of this inflationary period, imagine a balloon that is rapidly inflated, with the galaxies painted on the surface. The universe that we see populated by the stars and galaxies all lies on the surface of this balloon. Now draw a microscopic circle on the balloon. This circle represents the visible universe, everything we can see with our telescopes."

Alan Guth elaborates on his hyperexpansion Eureka moment in his book *The Inflationary Universe* [ISBN 0-201-32840-2] - when he describes the reaction after feeling relief at the favorable reception given to his inflationary theory lecture:

"By my count seventeen of the thirty-six lectures at the workshop were mainly about the idea of an inflationary Universe, and many of the other talks were heavily influenced by this idea. By supposing that the Universe inflates by an enormous factor, we explain at one stroke why it is observed to be so nearly flat, so nearly homogeneous, and perhaps why it is so nearly free of various topological monstrosities such as magnetic monopoles, axion domain walls, primordial black holes ... The idea is so simple, and yet it provides a qualitative understanding of some of the deepest puzzles of cosmology!"

Skipping back to Michio Kaku's book *Parallel Worlds*, we find a discussion of more recent ideas that build on the basic big bang theory:

"But recently, the tide has turned dramatically, with the finest minds on the planet working furiously on the subject. The reason for this sudden change is the arrival of a new theory, string theory, and its latest version, M-theory, which promises not only to unravel the nature of the multiverse but also to allow us to "read the Mind of God," as Einstein once eloquently put it....

String theory and M-theory are based on the simple and elegant idea that the bewildering variety of subatomic particles making up the universe are similar to the notes one can play on a violin string, or on a membrane such as a drum head. (These are no ordinary strings and membranes; they exist in ten and eleven-dimensional hyperspace.) ...

However, M-theory also features membranes; it is possible to view our entire universe as a membrane floating in a much larger universe. As a result, not all of these higher dimensions have to be wrapped up in a ball. Some of them, in fact, can be huge, infinite in extent.

One physicist who has tried to exploit this new picture of the universe is Lisa Randall of Harvard. Resembling the actress, Jodie Foster a bit, Randall seems out of place in the fiercely competitive, testosterone-driven, intensively male profession of theoretical physics."

Lisa Randall has written extensively on string theory and M-theory in her own book entitled *Warped Passages* [IBSN-13:978-0-06-053108-9]. In Chapter 3 she gives this example of a shower curtain as an example of 'branes' (short for membranes):

"Just as the water droplets on the curtain are bound to a two-dimensional surface, particles or strings can be confined to a three-dimensional brane sitting inside a higher-dimensional world. But unlike the drops on the curtain, they are truly trapped."

Expanding Universe In Scriptures?

So why are we going through all this back ground in order to begin discussing the opinions of God about the state of the universe? What bearing do hyper expansion, rubber balloons, shower curtains, and flexible membranes have on God's opinions? Perhaps, for some, it's like Yogi Berra's famous quotation "Its déjà vu all over again." We now go to the book, *The Dove Code* [ISBN 978-158169-2990] by Chatan N. Kallah (Chapter 15). The background of the story is that two college students, Sarah and Kim, are in an earnest discussion with Floyd—who believes that the creation account has many errors. They have been reviewing their ongoing courtroom debate about the sequence of creation / formation as given in Genesis. The subject of the inflation of the universe arises. We pick up the narrative as follows:

****

""I guess it's possible, but I don't think it's probable," responded Floyd.

"Oh, come on now, Floyd, remember in physics 435, we discussed inflation, hyperinflation, and the resulting flatness of the universe problems. Some people didn't like these concepts, but strangely enough the mathematical equations work beautifully when inflation factors were added. You remember those guys who worked on it – DeSittter, Guth and others?" asked Kim.

"Ok, I remember, but now I suppose you're going to tell me God created the universe by inflation?"

"Floyd, you're a perfect straight man. I couldn't have said it better myself. Matter of fact, we told you yesterday that inflation occurred during creation. You just didn't catch it." noted Kim.

"If you said anything about inflation yesterday, I certainly don't remember it."

"Floyd, inflation = stretched – remember? Let me borrow your laptop, Sarah. Ah, here it is."

JOB 26:7 **He stretcheth out** *the north over the empty place, and hangeth the earth upon nothing.*

"Oh, come on girls," said Floyd. "With all the stuff in the Bible, you pick out some little something here and there and say its proof. Well, in my teen years, I studied the Bible, too, and I know it says everything must be proven from the mouths of two or three witnesses. So, one little skimpy verse that only half says what you're trying to say kind of stretches, or should I say inflates the truth."

"Ok, you want more proof. Let me do a search on stretching and heavens," said Kim. "How about these?"

PSALMS 104:1-2 *Bless the LORD, O my soul. O LORD my God, thou art very great; thou art clothed with honour and majesty. Who coverest thyself with light as with a garment:* **who stretchest out the heavens like a curtain:**

"Or, how about this one?" asked Kim.

ISAIAH 40:21-22 *Have ye not known? have ye not heard? hath it not been told you from the beginning? have ye not understood from the foundations of the earth? It is he that sitteth upon the* **circle of the earth**, *and the inhabitants thereof are as grasshoppers;* **that stretcheth out the heavens as a curtain, and spreadeth them out as a tent to dwell in.**

"These two references speak of stretching out the heavens like a curtain. Doesn't this resemble the membrane theory we studied in our advanced physics class. Membranes can stretch, you know. I remember it because Lisa Randall, a mere woman, worked on its development. I heard her speak one time at a symposium," said Kim. "And you talk about Galileo and the flat earth—how about verse 22, where God is sitting on the circle, note it says *circle of the earth*? And here are more scriptures about stretching out the heavens. I'll keep going until you get off your '*picking out one verse to prove something*' high horse.'"

ISAIAH 42:5 *Thus saith God the LORD, he that created the heavens,* **and stretched them out; he that spread forth the earth,** *and that which cometh out of it; he that giveth breath unto the people upon it, and spirit to them that walk therein:*

ISAIAH 44:24 *Thus saith the LORD, thy redeemer, and he that formed thee from the womb, I am the LORD that maketh all things;* **that stretcheth forth the heavens alone; that spreadeth abroad the earth** *by myself;*

ISAIAH 45:12 *I have made the earth, and created man upon it: I, even my hands,* **have stretched out the heavens,** *and all their host have I commanded.*

Expanding Universe In Scriptures?

ISAIAH 51:12 I, even I, am he that comforteth you: who art thou, that thou shouldest be afraid of a man that shall die, and of the son of man which shall be made as grass;

13 And forgettest the LORD thy maker, **that hath stretched forth the heavens***, and laid the foundations of the earth; and hast feared continually every day because of the fury of the oppressor, as if he were ready to destroy? and where is the fury of the oppressor?*

JEREMIAH 10:12 He hath made the earth by his power, he hath established the world by his wisdom, **and hath stretched out the heavens** *by his discretion.*

JEREMIAH 51:15 He hath made the earth by his power, he hath established the world by his wisdom, **and hath stretched out the heaven** *by his understanding.*

ZEC 12:1 The burden of the word of the LORD for Israel, saith the LORD, **which stretcheth forth the heavens,** *and layeth the foundation of the earth, and formeth the spirit of man within him.*

"Now, let's see, we have verses from Job, David the writer of Psalms, Isaiah, Jeremiah, and Zechariah— five witnesses—is that enough for you?" asked Kim.

"Kim, you can quote all the scriptures you like, but the creation sequence just doesn't agree with our scientific theories."

"Well, Floyd, maybe your 'scientific theories' have a little catching up to do before you begin to fathom how God created the universe." replied Kim.

****

At the beginning this question was asked: Is the universe expanding – or in a steady state? Quotes were given examining the opinions of Einstein, Hoyle, and hopefully—if the scriptures have been interpreted right—the opinion of God. Now, perhaps you are forming your own opinion—that is unless you have previously made up your mind.

Scientists are now delving into questions about how the universe will end. Will it be the "big freeze" or the "big splat" or something else? The Bible does show a configuration of the universe and it's candlestick trunk and branch variables printed and displayed on the surface of a scroll: *And all the host of heaven shall be dissolved, and the heavens shall be **rolled together as a scroll**: and all their host shall fall down, as the leaf falleth off from the vine, and as a falling fig from the fig tree.* Could this be like a spring loaded window shade (a curtain) that is suddenly released and it rapidly rolls up?

Have fig leaves and figs from the Garden of Eden come to fruition and harvest? *And the heaven departed as **a scroll when it is rolled together**; and every mountain and island were moved out of their places.* Would a *new earth* have its mountains and islands tectonic plates rearranged? Can mountain and sea configurations really be changed by understanding the invisible mechanics of faith?

Jesus answered and said unto them, Verily I say unto you, If ye have faith, and doubt not, ye shall not only do this which is done to the fig tree, but also if ye shall say unto this mountain, Be thou removed, and be thou cast into the sea; it shall be done.

It is interesting to speculate about membranes, curtains, scrolls, and whirling cylindrical black holes, but at this point, 'Wait and see' seems to be the best approach as the unseen dimensions are becoming seen. However, will those with childlike faith be the first to 'see' and enter this new realm?

- ♦ Context Scripture Chapters: *Psa 139, Job 26, Psa 104, Isa 40&44&45&51, Jer 10&51, Zec 12, Isa 34, Rev 6&21, Mat 21*

𝔄rticle 𝔖ix

THE LORD'S TIME TRAVEL PARADOX

II Peter 3:8
"…one day is with the Lord as a thousand years"

Why does this scripture relate one day to a thousand years? Why not have one day as one hundred years? Or, one day as a million years?—or maybe one day as two billion years? The answer may surprise you. Is there any scientific basis for the relationship described in II Peter 3:8?

Many of today's well known astrophysics authors have given examples about time travel where one gets on a rocket and speeds off to a distant point at some high fraction of the speed of light and then returns to a surprise. Stephen Hawking in Chapter 6 of his book, *A Briefer History of Time* [ISBN-13:968-0-553-80436-2] discusses the twins paradox: "Our biological clocks are equally affected by these changes in the flow of time. Consider a pair of twins … if one of the twins went for a long trip in a spaceship in which he accelerated to nearly the speed of light. When he returned, he would be much younger than the one who stayed on earth."

So, what would be a finite example of God traveling in a spaceship be —as we endeavor to broaden our understanding of the infinity of God? How fast would the spaceship have to go in order for one day with God to be a thousand years on earth? We know that the speed of light is 186,000 miles/second. While we understand telephone and satellite communications occurring almost instantly, it still seems foreign to our understanding for a human body to be transported that fast. So, we will imagine that we are in a Ferrari with a speedometer indicator showing a top speed of 200 mph with a mysterious **RED ZONE** starting at 186 mph

(However, behind the scenes, we understand that the red zone really starts at 186,000 miles per second).

If we accelerate our Ferrari to the point that one day in our space car is equivalent to 1000 years on earth, how many earth days would that be?

1000 years X 365.24 days/year = 365,240 days

Fortunately, on our instrument panel, we will also have a trip calculator, that tells us how many days have elapsed on earth while we are accelerating in our space car. We have been told this trip calculator uses a neat little formula known as the Lorentz factor to calculate the ratio of days on earth to days in our space car. This factor is 1.0 divided by the square root of [1.0 minus (actual velocity squared divided by the speed of light squared)]. However, since our neat little trip computer does all this for us, we will not unsafely get bogged down by making detailed calculations while driving.

So, folks, fasten your seat belts, rev up the engine and let's accelerate in twenty mile per hour increments and see what our instrument panel tells us. What happens when the indicator reaches the **RED ZONE**?

We have accelerated up to 100 and there is a small time difference of 1.186 days.

Speedometer	0	20	40	60	80	100
Earth Days	1	1	1.024	1.056	1.108	1.186

Now, let's put the pedal to the metal and see what this baby will do!

Speedometer	100	120	140	160	180	200
Earth Days	1.186	1.309	1.519	1.961	3.969	RED!

Whoops! When we hit a little over 180, we seemed to have stalled out. Is this something like the sound barrier that Chuck Yeager kept running up against? However, at the 180 speed we did get up to having almost four days [3.969 days] pass on earth while we would have aged only one day. Let's take a different tack. We will accelerate up to 180 and then gradually increase our speed to creep up on what is actually happening. Instead of a 'lead foot' on the accelerator, we will need a 'feather foot' – otherwise we will **RED ZONE** our Ferrari.

However, our 'Nerd' friends advise us that we will need a very special speedometer to approach this light barrier. They have provided a speedometer that 'zooms in' just under the 186.0 mph mark and will have twelve decimal points to register speeds as high as 185.999999999999 mph.

The math whizzes laugh at us for having to use 186 miles per hour in an earthbound Ferrari as an example to represent 186,000 miles per second light speed. However, if these Nerds would leave their computer screens for a few hours and enjoy a drive out in the bright sunshine amongst the trees and flowers in God's green earth, they might have more of a social life. But they laugh at us and say that they have already 'mathematically' broken the 186,000 mile per second 'light barrier' and are 'outside' of time and space — because time and space have merged into a 'singularity'.

Ok, so much for 'math speak', we have enlisted the services of a young lady, with a very delicate touch on the accelerator, to help us ease up on the 186.0 barrier and see what happens. Here we go! Whee! Easy now—we're at 180—now at 185—and easing up on 186.

Speed	180	185	186	186	185.9999	185.99999999
Days	4	10	31	96	964	96,436

Now our space car has really accelerated—in one day our earth bound friends have aged 96,436 days (265 years). Chuck Yeager, here we come!

Speed	185.99999999	185.999999999299	185.999999999999	186.0
Days	96,436	364,238	9,686,330	RED!

Wow! When we got to a speed of **185.999999999299 K miles per sec,** we were just under the 364,240 days in 1000 years of earth time. But, when we try to nudge it a little further, the number of earth days gets up in the millions and it finally blows up to an infinite number at the light barrier. Similar to the sound barrier, practical considerations overwhelm math-speak theory.

But, if a thousand years of earth time is equivalent to one day with the Lord, is the Lord limited by the light barrier, too? A good question, but we wonder if the light barrier is something that so far has limited mankind and not God? Perhaps, the limitations of this old earth will be done away with when the new heaven and earth that the Apostle Peter wrote about comes to fruition: *Looking for and hasting unto the coming of the day of God, wherein the heavens being on fire shall be dissolved, and the elements shall melt with fervent*

The Lord's Time Travel Paradox

heat? *Nevertheless we, according to his promise, look for new heavens and a new earth, wherein dwelleth righteousness.*

You have heard it said that God exists outside of time. But how could He exist outside of time? Astrophysicists are fond of talking about the 'big bang.' We pose a question: How much time would pass for God if He initiated the big bang and then traveled as light outward from its center? Well, if God were light, we can see by the above speedometer examples that **zero** time would transpire for God, while a very large amount of time would take place in earth time. How could God be light? The Apostle John wrote these key statements concerning God and Jesus: *This then is the message which we have heard of him, and declare unto you, that God is light, and in him is no darkness at all. Then spake Jesus again unto them, saying, I am the light of the world: he that followeth me shall not walk in darkness, but shall have the light of life. I am Alpha and Omega, the beginning and the ending, saith the Lord, which is, and which was, and which is to come, the Almighty.* Are we to take John at his word and say that God existed before time as we know it, exists now, and already exists in our future when a new heaven and earth comes forth?

Paul Davies, Arizona State University professor, gives a description of space-time designer-creator relationships in his book: *The Goldilocks Enigma* [ISBN 13: 978 0547053585]. Why Is the Universe Just Right for Life?

"A Cosmic Designer Must Lie Outside Time

There is also the very considerable problem of time. Time is part of the physical universe, inseparable from space and matter. Any designer-creator of the universe must therefore transcend time, as well as space and matter. That is, God must lie *outside* time if God is to be the designer and creator *of* time. Augustine was well aware of this and began a school of thought that asserts that God is a timeless being, not just in the sense of living forever but of being outside time altogether." [Page 200]

Is God limited by the light barrier—or is this just a barrier that exists in the minds of men? Michio Kaku in his book entitled **Einstein's Cosmos** [ISBN 0-393-05165-X] relates how Einstein struggled with this because some experiments were indicating that twin electrons, when separated at large distances from each other were communicating at speeds faster than the speed of light. In Chapter 7, Michio writes: "Now suppose that you finally measure the spin of one electron. It is for example, found to be spinning up. Then instantly, you know the spin of the other electron, although it is many light-years away—since its spin is the opposite of its partner, it must be spinning down. This means that a measurement in one part of the universe instantly determined the state of an electron on the other side of the universe, seemingly in violation of special relativity. Einstein called this "spooky action-at-a-distance." ... Einstein disliked this idea, because it meant that *the universe was nonlocal*; that is, events here on Earth instantly affect events on the other side of the universe, traveling faster than light."

Could this be what King David described when he wrote in the Psalms: *O LORD, thou hast searched me, and known me. Thou knowest my downsitting and mine uprising,* **thou understandest my thought afar off**. *... Whither shall I go from thy spirit? or whither shall I flee from thy presence? If I ascend up into heaven, thou art there: if I make my bed in hell, behold, thou art there. ... If I say, Surely the darkness shall cover me; even the night shall be light about me.* Or what did the Apostle Paul mean when he said this about the heroes of faith? *These all died in faith, not having received the promises,* **but having seen them afar off**, *and were persuaded of them, and embraced them, and confessed that they were* **strangers and pilgrims on the earth**.

If God understands our thought "afar off" then perhaps when the Chuck Yeager's among us break through the light barrier, we will break free into the world of thought. However, David said that even when he tried to hide in the darkness, *even the night shall be light about me*. And apparently that light even extends into heaven and hell (Parallel Universes?).

But how could God be light—wouldn't that mean He would have to be everywhere at once? Clearly, there have been certain manifestations of God where He lowered Himself in frequency to give the Ten Commandments to Moses and where Jesus stripped himself of his heavenly glory to walk among men. Imagine what would happen if we directly wired a 220,000 volt electrical transmission line directly into a 220/110 volt household without having a transformer to step down the voltage? Or, consider that man's definition of 'light' is limited to the visible range between infrared and ultraviolet. This is a very, very, tiny sliver of the electromagnetic spectrum – man's eye is oblivious to the rest of the spectrum. No wonder the scriptures speak *of blind guides, which strain at a gnat, and swallow a camel*.

The Greeks loved reason and logic. The Apostle Paul advised them of God being everywhere in this fashion: *For as I passed by, and beheld your devotions, I found an altar with this inscription, TO THE UNKNOWN GOD. Whom therefore ye ignorantly worship, him declare I unto you. ... they should seek the Lord, if haply they might feel after him, and find him, though he be not far from every one of us: For in him we live, and move, and have our being; as certain also of your own poets have said, For we are also his offspring.*

But what of the creation controversy—did God create the earth in six literal twenty four hour days and then rest on the seventh day? This would be a good question to ask of one who is not subject to the limitations of time. We discussed the "twins' paradox" as described by Stephen Hawking. There are number of similar examples given by other writers where someone hops in a spaceship, approaches the speed of light, and then returns to be shocked that so much time has passed. In the book, *The Dove Code* [ISBN 978-158169-2990], by Chatan N. Kallah, a similar example is given in Chapter 11 about Methuselah's nephew—by CowboyBobR.com. CowboyBobr discusses the elasticity of time, and that how much time passes simply depends on one's 'relative' viewpoint. In the book, two college students, Kim and Sarah theorize what happen if the passing of time was greatly hyper expanded. See their example from Chapter 11 about 'Fast Eddy's dilemma.

"CowboyBobr went on to spin a yarn about Methuselah's nephew who decided to sneak his dad's rocket ship out for a spin—thinking his folks wouldn't know because they were away celebrating an anniversary in Paris. Being like most teenage boys, he wanted to see what kind of speed he could get out of the sleek baby and he pulled the control stick back to full warp. Unfortunately, it took him some time to determine how to unwarp the rocket to get back to home base before his dad discovered his escapade. When he got back, he was in total shock to find himself attending his Uncle Methuselah's 969th birthday party!

After getting over his shock, he was told that his mother, Dad and siblings had long, long passed away and only his Uncle Methuselah was still living. Methuselah credited his long life to what his dad Enoch had taught him in his younger years, but, as he aged, his memory of his dad's teachings dimmed.

Who was right? Did a lot of time pass as evidenced by Uncle Methuselah's age or did a short time pass as evidenced by the nephew taking the ship out for a spin and trying to get back before he thought his Mom and Dad would return? CowboyBobr said that every story should have a moral: *One's opinion of whether the* **ELAPSED TIME** *is long or short depends on which side of hyperexpansion [or hypercontraction] one is considering it. It is all relative you know.*

Kim was intrigued that the email in CowboyBobr's Methuselah example was actually more in line with modern man's view from the earth of classic hypercontraction rather than hyperexpansion. So, she and Sarah decided to put their heads together and come up with a hyperexpansion example.

It seems that Methuselah's nephew Fast Eddy wanted to impress his girlfriend by taking her for a spin in his dad's rocket while his parents were away in Paris. As they left metropolitan space into the countryside, his girl friend cuddled up next to him and began giving him amorous kisses. Not wanting to have distractions from his priorities, Fast Eddy put the rocket in **autopilot_time_lapse_hyperexpansion** and proceeded to return his gal's kisses. It was like having a honeymoon on a slow boat to the China galaxy—the longer it lasted—the better. However, after a time, Fast Eddy wanted to get back before his parents returned and his gal's folks missed her. When he tried to reverse the elapsed time hyperexpansion, there was no response; the autopilot relay was stuck in the 'on' position. Many years later, Fast Eddy found the wiring diagram under the dash and devised a bypass. The return home was a shocker to Eddy's parents. They had only been in Paris for seven days and when they returned, they found they had a new daughter-in-law and were grandparents, great grandparents, and great, great grandparents as all 64 members of Eddy's family returned to greet them. Eddy had a lot of explaining to do."

Well, Fast Eddy's story is perhaps a little on the humorous side, but it does illustrate there is a lot about time that we don't yet understand. Did Fast Eddy's trip take 'seven days' or seventy years? Perhaps, if we did understand time, it would 'blow our minds.' But, could that be a good thing? Consider how Job's mind was blown when God began to reveal the unfolding of creation to him. In modern day terms, we would say that God took Job to the singularity at the vortex of the whirlwind of creation — a place where space and time were no longer separate. We go to the account in the book of Job: *Then the LORD answered Job out of the whirlwind, and said, Who is this that darkeneth counsel by words without knowledge? Gird up now thy loins like a man; for I will demand of thee, and answer thou me.*

Stephen Hawking described a similar scenario at the conclusion of his *A Brief History of Time* book [ISBN-13:978-0-553-80436-2]:

"Even if the whole universe did not re-collapse, there would be singularities in any localized region to form black holes. These singularities would be an end of time for anyone who fell into the black hole. At the big bang and other singularities, all the laws would have broken down, so God will still have had complete freedom to choose what happened and how the universe began."

So, imagine that out of the whirlwind—God were asking you these questions instead of asking Job.

Where wast thou when I laid the foundations of the earth? *declare, if thou hast understanding. Who hath laid the measures thereof, if thou knowest? or who hath stretched the line upon it? Whereupon are the foundations thereof fastened? or who laid the corner stone thereof; When the morning stars sang together, and all the sons of God shouted for joy?*

These would be tough questions for anyone to answer. However, there is something really peculiar about the book of Job. It is one of the older books of the Bible thought to have been written in the range of about 500 years before the birth of Christ. Approximately 2000 years later, Galileo got into his little tift with the so-called 'official' doctrine of church clergy. But, in fairness to the parties involved, let's consider the scene that was described in Job Chapter 26. Curiously, one of the more recent theories about the formation of the universe involves hyperexpansion of a stretchable 'membrane' and has been dubbed 'M-theory.'

He stretcheth out **the north over the empty place, and** *hangeth the earth upon nothing.* **He bindeth up the waters in his thick clouds; and the cloud is not rent under them.**

He holdeth back the face of his throne, and spreadeth his cloud upon it.

He hath compassed the waters with bounds, until the day and night come to an end.

When the Apollo astronauts visited the moon and looked back on the earth, could they have described the view any more perfectly? They saw the earth suspended in space, the white clouds over the blue waters. They saw the dividing line between night and day. And then, contrary to the 'flat earth' theory they saw the compassing (a circle) of the waters of the spherical earth. How could Job have written such a description long before the space age? We add this to the inbox of mysteries that we have already discussed. It blows one's mind. But, perhaps it needs to be blown so that the mind of Christ can meld with our mind.

- Context Scripture Chapters: *IIPet 3, Joh 1, Rev 1, Psa 139, Heb 11, Act 17, Job 38&26*

Article Seven

DO LIFE FORMS EXIST AMONG THE STARS?

Why Can't Religion and Science Agree?

Where did the human race come from? Did it arise from some one celled creature that evolved out of a primordial soup? Many, but not all, in the scientific realm think so. However, creationists have opposed this view saying that man's beginnings are as recorded in the book of Genesis—ascribing man's formation to the handiwork of God. Others have tried to bridge the gap by saying that God used evolution as a process in the creation of man. Both sides seem to agree that man's body contains the elements or *the dust of the ground.* But, did life arise randomly and spontaneously, or did the DNA blueprint for man come from somewhere else? Is that blueprint part of the *"precious seed"* described in the Bible? Even those scientists who stoutly defend evolution — when pressed—are reluctant to rule out that seeds of life could have been brought to earth from sources or even from civilizations in another part of the universe.

And then, there are more "*afar off*" theories that quote the Apostle Paul's writing that the heroes of the faith were given promises. These faithful having seen the promises "*afar off, and were persuaded of them, and embraced them, and confessed that they were strangers and pilgrims on the earth.*" But what are other civilizations and how does God fit into this puzzle? College students particularly struggle with these concepts as they compare their parent's beliefs with what is currently presented in the majority of the universities.

Many remember Carl Sagan as host on *Cosmos: A Personal Voyage*— the most widely watched PBS program in history reaching over 600 million people in 60 countries. He promoted and co-pioneered SETI—the Search for Extra-Terrestrial Intelligence. He was intensely curious about finding other civilizations *out there* and famous for his pronunciations about the _bee_lions upon _bee_lions of stars in the universe. He summarized his concept of God in this statement:

"The idea that God is an oversized white male with a flowing beard, who sits in the sky and tallies the fall of every sparrow is ludicrous. But if by 'God' one means the set of physical laws that govern the universe, then clearly there is such a God. This God is emotionally unsatisfying ... it does not make much sense to pray to the law of gravity."

Sagan and others worked on developing a classification system for civilizations that might be found in this solar system and beyond to the "billions upon billions" of stars in the universe. Given that the number of stars is large—possibly beyond comprehension—it is thought that a high probability exists of finding civilizations like earth [or perhaps civilizations millions of years more advanced] somewhere "*afar off*" in space.

However, Sagan's concept of the infinity of the stars and the Apostle Paul's writing were not all that much different. The Apostle Paul wrote about a seed that would multiply to be "*so many as the stars of the sky in multitude, and as the sand which is by the sea shore innumerable.*"

Sagan did not live to see his book *Contact* made into the movie that was released in 1997, but he did leave a legacy for those who would probe the universe. Part of that legacy is Michio Kaku who entered into the TV limelight as an interpreter of theoretical physics and has written numerous books on the subject. Kaku graduated summa cum laude from Harvard University with a B.S. degree in 1968 and was first in his physics class. He attended the University of California, Berkeley, and received a Ph.D. in 1972. He discusses his upbringing in chapter one of his book *Parallel Worlds* [ISBN 0-385-50986-3] – a book that probes the possibility of a universe or universes parallel to our own.

"When I was a child, I had a personal conflict over my beliefs. My parents were raised in the Buddhist tradition. But I attended Sunday school every week, where I loved hearing the biblical stories about whales, arks, pillars of salt, ribs, and apples. I was fascinated by these Old Testament parables, which were my favorite part of Sunday school. It seemed to me that the parables about great floods, burning bushes, and parting waters were so much more exciting than Buddhist chanting and meditation. In fact, these ancient tales of heroism and tragedy vividly illustrated deep moral and ethical lessons which have stayed with me all my life. ...

Today, however, a resolution seems to be emerging from an entirely new direction—the world of science—as the result of a new generation of powerful scientific instruments soaring through outer space. Ancient mythology relied upon the wisdom of storytellers to expound on the origins of our world. Today, scientists are unleashing a battery of space satellites, lasers, gravity wave detectors, interferometers, high-speed supercomputers, and the Internet, in the process revolutionizing our understanding of the universe, and giving us the most compelling description yet of its creation."

In chapter 13 of Kaku's book *Hyperspace* [ISBN 0-385-47705-8], he discusses futurology and types of civilizations that might arise and also their mastery of the additional dimensions proposed in string theory.

"Futurology, or the prediction of the future from reasonable scientific judgments, is a risky science. Some would not even call it a science at all, but something that more resembles hocus pocus or witchcraft. Futurology has deservedly earned this unsavory reputation because every "scientific" poll conducted by futurologists about the next decade has proved too wildly off the mark. What makes futurology such a primitive science is that our brains think linearly, while knowledge progresses exponentially. ...

With all these important caveats, let us determine when a civilization (either our own or possibly one in outer space) may attain the ability to master the tenth dimension. Astronomer Nikolai Kardashev of the former Soviet Union once categorized future civilizations in the following way. ...

The basis of this classification is rather simple: Each level is categorized on the basis of the power source that energizes the civilization. Type I civilizations use the power of an entire planet. Type II civilizations use the power of an entire star. Type III civilizations use the power of an entire galaxy. This classification ignores any predictions concerning the detailed nature of future civilizations (which are bound to be wrong) and instead focuses on aspects that can reasonably be understood by the law of physics, such as energy supply.

Our civilization, by contrast, can be categorized as a Type 0 civilization, one that is just beginning to tap planetary resources, but does not have the technology and resources to control them. A Type 0 civilization like ours derives its energy from fossil fuels like oil and coal, and, in much of the Third World, from raw human labor. Our largest computers cannot even predict the weather, let alone control it. Viewed from this larger perspective, we as a civilization are like a newborn infant."

In order for an infant to be birthed, it must go through a constriction from the womb into a new world of light and knowledge. The recently released book, *The Dove Code* [IBSN 978-158-169-2990], by Chatan N. Kallah discusses such a constriction where both scientists and creationists are being forced into a birth canal that may have the light of a new millennium at the end of the tunnel. The setting is a university prelaw school courtroom drama where two students, Sarah and Kim are debating future lawyers Floyd and Harry about whether the Bible's representation of God making the sun on the fourth day is scientifically plausible. Floyd's major witness, the internationally known astrophysicist, Dr. Bertrand Eisenhoff, has just completed laying out a presentation of how the sun was first formed and the planets began their orbits. We pick up the book's narrative on page 187 where Kim begins her questioning of Dr. Eisenhoff, a science advisor to world leaders, who has made time in his busy schedule for this university prelaw courtroom debate.

""Dr. Eisenhoff, thank you for showing us such a beautiful and well done presentation on the stars and the formation of the planets."
"You're welcome," brightened Dr. Eisenhoff. "It was my pleasure."
"I have a few questions for you," said Kim. "The first is that after watching the supreme beauty of galaxies and the universe as shown in your videos, how can you not believe there is a God?"
Harry Jamieson, the court protocol advisor, gave Floyd a sharp nudge in the ribs.
"Objection, Your Honor! The witness' belief or lack of belief in God is not on trial here," protested Floyd.
"Objection granted," said the judge.
"Dr. Eisenhoff, what are your views about the Bible?" asked Kim.

"Objection! Your Honor, " cried Floyd. "Dr. Eisenhoff is here as a scientific witness and not here to give his personal views about the Bible."

Dr. Eisenhoff looked at Floyd and then the judge and said, "Your Honor, I would be happy to answer the question and give my views about the Bible."

The Judge looked at Floyd who nodded OK. "I'll allow it. You may proceed to give your answer, Dr. Eisenhoff."

Dr. Eisenhoff began, "I would treat the Bible just like any other history book. History books have errors in them and so does the Bible. In my opinion, some of the stories in the Bible are clearly myths that were made up in the imaginations of the writers. Some of the so-called miracles are not scientifically plausible."

Dr. Eisenhoff continued, "Yes, the teachings of Moses and Jesus are good moral standards to live by, but you find most of these same moral standards in other religions. I respect the Bible as a source of history, literature, and moral values, but I do not believe it to be sacred. And I certainly believe that passages in the Bible should not be used to inhibit the advancement of scientific knowledge."

"Thank you for graciously answering the question, Dr. Eisenhoff. I think it was important for the jury to understand your personal viewpoint of the Bible and the effect, if any, your viewpoint might have on your scientific conclusions," said Kim.

"You're welcome," smiled Dr. Eisenhoff, thinking that it was fortuitous that he was allowed to say the things he knew should be said about the Bible.

"I have several other questions for you," continued Kim. "Given the billions and billions of stars in the universe, do you think there might be other planets where life forms comparable to earth exist?"

"Yes, when you do mathematical probability analyses, given the tremendous number of stars in the universe, it does appear life forms may exist elsewhere because they exist here on earth."

"Are you familiar with the classes of civilizations some of your colleagues have proposed that may exist at various places in the universe?" asked Kim.

"I certainly am," responded Dr. Eisenhoff. "I participated in setting up these models."

Sarah noticed that Harry Jamieson was conferring in a nervous whisper with Floyd. She hoped Kim would be able to continue her dialog just a little longer before the line of questioning received an objection from the plaintiff. However, Dr. Eisenhoff seemed to be enjoying showing off his vast knowledge.

"Would you tell us what these classes of civilization are?" asked Kim.

"They are numbered one through four, with four being the highest level of civilization. The earth is approaching a level one civilization, and when we can control weather and develop certain other technologies, we will have reached level one."

"What about level four?" asked Kim, "What are its characteristics?"

"Oh, level four is the very highest technology," replied Dr. Eisenhoff. "Travel through the universe will be routine. Problems with food, weather, and environment will completely be problems of the past. It will be a utopia for mankind."

Kim asked, "Dr. Eisenhoff, have you ever considered the possibility the level four civilization might be God and His holy angels?"

A stunned look spread over Dr. Eisenhoff's face. "Uh, no, I never really gave that any thought."

There was a ripple through out the courtroom. Floyd jumped to his feet and roared, "Your Honor, Your Honor, I move the last question and the answer given be stricken from the court record! The counselor is leading the witness!"

"Motion granted," ordered Judge Emerson. "The last question and its answer are to be stricken from the court record. The jury is instructed to disregard the question and its answer. The counselor is instructed to refrain from leading the witness.'"

So the debate between the scientists and creationists rages on. But where did man come from? If it were not so serious, it would be almost comical. Scientists willingly acknowledge the probability of more advanced civilizations — but many strain mightily to avoid using the 'God' word. In the DNA discovery—post Darwin discovery period, the improbability (or the probability) of random assembly of complex DNA structures has become a major sticking point. On the other hand religion often strains mightily to avoid ideas challenging their traditions – some of which are clearly man based rather than God based. Do life forms exist among the stars? Scientists and religionists do—kind of—sort of—maybe—could possibly—agree that they do exist.

Both scientists and theologians are currently in an odd position—like the audience at a theatre waiting for the third act of the play to begin.

The first two acts have essentially in the words of the old saying—kicked the can down the road—waiting for a later resolution. For example, Einstein thought the universe was static rather than expanding. However, numerous scriptures speak of the process *that stretcheth out the heavens as a curtain*. Later, Hubble and others gained enough knowledge to confirm the stretching out and later the membrane (*a curtain?*) theory was developed. The knowledge of scientists is not yet complete enough to understand faith along with the scriptures emanating from the invisible dimensions of the spirit world. And theologians have the invisible faith of the unseen spirit world, but have not yet been able to birth answers in terms of worldly parameters. The third act written, as the scriptures say, by *the author and finisher of our faith*, is awaited with eager anticipation.

The Bible says *"And the LORD God formed man of the dust of the ground, and breathed into his nostrils the breath of life; and man became a living soul."* But where did the dust of the earth come from and was there an overall blueprint for man's design that was used by a Creator?

Perhaps, we need the Wisdom of Solomon when he quoted Wisdom in the book of Proverbs as saying: "*The LORD possessed me in the beginning of his way, before his works of old. I was set up from everlasting, from the beginning, or ever the earth was. When there were no depths, I was brought forth; when there were no fountains abounding with water. Before the mountains were settled, before the hills was I brought forth:* **While as yet he had not made the earth, nor the fields, nor the highest part of the dust of the world."**

- Context Scripture Chapters: *Gen 1, Psa 126, Heb 11, Gen 2, Pro 8*

Article Eight

IF I TRY AND BEND THAT FAR, I'LL BREAK

The Truth Shall Make You Free

Each day upon the earth, as the sun rises, millions go about their daily lives providing bread for the table, nurturing their children, laughing, crying and participating in our civilization's activities. If like on a Google map, we could press a zoom button and take in the perspective of time and space from a distance what would we see? Suppose we look over the shoulder of a NASA artist and describe the steps as he touches the paint to a canvas to portray the scene laid out before us.

He stretches out the north over the empty place, ***and hangs the earth upon nothing****. He binds up the waters in his* ***thick clouds****; and the cloud is not rent under them. He holds back the face of his throne,* ***and spreads his cloud upon it. He has compassed the waters with bounds, until the day and night come to an end.***

This creation scene of the spherical earth hanging upon nothing was very artistically described by Job in the 26th chapter of one of the oldest books of the Bible. It transcends a host of traditional views that were embraced by both scientists and theologians over the earlier years about the nature of the earth and its place in the universe.

Well, what is a traditional view? It is something that is generally embraced by our culture. And, if it is based on truth, it is a helpful blessing. But, if it has some twists and turns in it that are not quite true—we can end up in a swamp of confusion.

You may remember the milkman singing about his puzzlements concerning tradition in the classic play and movie *Fiddler on the Roof.* This father was blessed with five daughters and it was traditional that the

If I Try And Bend That Far, I'll Break

parents would choose the future spouses for their daughters. However, a clash of wills arose when the oldest daughter wanted to choose a spouse she loved rather than the older and richer suitor that her parents and the matchmaker proposed. Eventually, love triumphed over tradition—but at least the bridegroom was of like faith as the oldest daughter.

Later the third daughter fell in love with a Christian boy and asked the father's blessing for the marriage. He refuses, she elopes, and the father considers the daughter as being dead because blessing an offspring's' marriage with one of another faith is a line he cannot cross.

Toward the end of the play, the expulsions of the Hebrews from Tsarist Russia in the early 1900's impacted the family and on short notice they packed up meager belongings in their wooden carts and left for Poland. As they were leaving, the 'dead' daughter came to say good bye to her family. The father refused to speak to his daughter, saying, "**If I try and bend that far, I'll break**!" But, as they walked away he instructed his wife to whisper, "God bless you."

Our evaluation of the role of tradition in this play may depend on whether we are Hebrew, Christian, or something else. We will set aside (for these writings) the classic Israelite - Christian confrontational arguments and agree that both are blessed. For the Apostle Paul wrote: *As concerning the gospel, they are enemies for your sakes: but as touching the election, they are beloved for the fathers' sakes.*

It would be safe to say that our viewpoint of the above is colored by the traditions which we have experienced. But, when is it OK to break with tradition and when is it best to stay within what our parents and / or our peer group is telling us is the tried and true way to go. The Apostle John wrote: *And ye shall know the truth, and the truth shall make you free.* Oh, if we could just wave a magic wand and know what is true. But, history tells us learning the truth is a process.

When it comes to traditions that are ultimately set aside, neither the 'church' nor 'science' has been devoid of problems. Classic confrontations took place with the established church during the Galileo and Copernicus years. Science, on the other hand, has a path of theories that were thought to be true but have now been discarded. But, perhaps the subject of the very creation of the sun, moon, stars and the earth is the most controversial. What traditions should be retained and what should be discarded? And what do we not yet know enough about in order to make a proper evaluation?

In the first article in this series entitled CHAIN OF CUSTODY OF CREATION – What Happened Before Genesis? – we developed a sequence that began before the earth had been made and culminated in the steps in Genesis chapters 1 and 2. This is based on wisdom's narrative about *the highest part of the dust of the world.*

PROVERBS 8:1 ... 31 *Doth not wisdom cry? and understanding put forth her voice? She standeth in the top of high places, by the way in the places of the paths. ...*

The LORD possessed me in the beginning of his way, before his works of old. I was set up from everlasting, **from the beginning, or ever the earth was.** *When there were no depths, I was brought forth; when there were no fountains abounding with water.*

Before the mountains were settled, before the hills was I brought forth: **While as yet he had not made the earth**, *nor the fields, nor* **the highest part of the dust of the world.** *When he prepared the heavens, I was there: when he set a compass upon the face of the depth:*

When he established the clouds above: when he strengthened the fountains of the deep: When he gave to the sea his decree, that the waters should not pass his commandment: when he appointed the foundations of the earth: Then I was by him, as one brought up with him: and I was daily his delight, rejoicing always before him; Rejoicing in the habitable part of his earth; and my delights were with the sons of men.

We continued in this evaluation to point out that the time that expired when the *dust of the world* existed and beginning verses of Genesis chapter 1 is not specified—therefore the elements of dust could have been thousands, millions, billions, or trillions of years old. We also presented concepts that the life forms on earth could have come from other sources in the universe rather than spontaneously sprung up from the earth itself. The Apostle Paul even wrote: *These all died in faith, not having received the promises, but having seen them afar off, and were persuaded of them, and embraced them,* **and confessed that they were strangers and pilgrims on the earth.** In that case, the timeline for life form development in another part of the universe is unknown. And then we summarized our article one findings as follows:

Granted there are many questions to be answered to fill in details, but in this article we have provided a framework for the creation of the earth and the life living upon it (later articles will address more details). And that framework presents a chain of custody from *the dust of the highest part of the world* to the earth itself and to the *dust of the ground* that was used to plant Adam and other life forms on the earth. And curiously enough, this framework does not depend upon the totality of the creation process being done in twenty-four hour days approximately 6000 years ago. For we do not know the period of time that the *dust* pre-existed or the seed pattern design pre-existed before the events of Genesis 1:1 Our geologist and paleologist friends will undoubtedly have both questions and contributions to make in fitting details into that framework—**that will be covered in a subsequent article in this series.**

Now, we have arrived at that subsequent article. What about all the geological layers and the times geologists have projected for their formation? What about the life forms that have been found within these layers? We will attempt to answer some of these questions by examining the first verses of Genesis.

GEN 1:1 *In the beginning God created the heaven and the earth. 2 And* **the earth was without form, and void** *[chaotic]; and* **darkness was upon the face of the deep**. *And the Spirit of God moved upon the face of the waters.*

In our first article, we have presented flashbacks from various parts of the Bible to Genesis period—indeed the quote from Proverbs chapter 8 is a flashback. Now, we will examine other verses and ask the question if these are flashbacks. Look at the following sequence of verses and ask yourself when evil first appeared among the sons of God. Most Christians believe that a rebellion among the sons of God happened at some period in the past. *Now is the judgment of this world: now shall the prince of this world be cast out*. Certainly evil had appeared by the time of the serpent's entry into the garden. Did the *prince of this world* (Satan) contribute to the *chaos* described in Genesis 1:2? Did Satan and his cohorts mess up the original creation joyfully brought forth by the sons of God?

When the foundation of the earth was laid, the Bible tells us *all the sons of God shouted for joy*. Then at some point, it was said *all the foundations of the earth are out of course*. Carefully examine the one through five statements below and ask yourself if they are in a timeline sequence.

The Articles of Configuration

1. JOB 38:4 **Where wast thou when I laid the foundations of the earth?** *declare, if thou hast understanding. ... 6 Whereupon are the foundations thereof fastened? or who laid the corner stone thereof; 7 When the morning stars sang together,* **and <u>all</u> the sons of God shouted for joy?**

2. PSA 82:1 *God standeth in the congregation of the mighty;* **<u>he judgeth among the gods</u>**. *2 How long will ye judge unjustly, and accept the persons of the wicked? Selah. 3 Defend the poor and fatherless: do justice to the afflicted and needy. 4 Deliver the poor and needy: rid them out of the hand of the wicked. 5* **They know not, neither will they understand; they walk on in <u>darkness</u>: all the <u>foundations of the earth are out of course</u>**. *6 I have said,* **<u>Ye are gods</u>**; *and all of you are children of the most High. 7 But* **ye shall die like men**, *and fall like one of the princes.*

3. JER 4:23 *I beheld the earth, and, lo,* **<u>it was without form</u>**, *and void; and* **<u>the heavens, and they had no light</u>**. *24 I beheld the mountains, and, lo, they trembled, and all the hills moved lightly. 25 I beheld, and, lo, there was no man, and all the birds of the heavens were fled. 26 I beheld, and, lo, the fruitful place was a wilderness, and* **<u>all the cities thereof were broken down at the presence of the LORD</u>**, *and by his fierce anger. 27 For thus hath the LORD said, The whole land shall be desolate; yet will I not make a full end.*

4. GEN 1:2 *And the* **<u>earth was without form, and void</u>**; *and* **<u>darkness was upon the face of the deep</u>**. *And the Spirit of God moved upon the face of the waters.*

5. GEN 3:22 *And the LORD God said, Behold, the man is become as* **<u>one of us, to know good and evil: and now</u>**, *lest he put forth his hand, and take also of the tree of life, and eat, and live for ever:*

If the above verses are arranged in sequence, then we have to deal with the proposition that life existed and cities existed and *were broken down*—as part of the chaos and darkness of Genesis 1:2. This would propose a pre-Adamite creation. Or, we could take a more traditional view and say that Psalm 82 and Jeremiah 4 are describing the events of Noah's flood. So, let's take a look at the description of the events of Noah's flood and compare them against the above sequence.

> GEN 6:1 *And it came to pass, when men began to multiply on the face of the earth, and daughters were born unto them,* 2 **That the sons of God saw the daughters of men that they were fair; and they took them wives of all which they chose.** *3 And the LORD said, My spirit shall not always strive with man, for that he also is flesh: yet his days shall be an hundred and twenty years. 4 There were giants in the earth in those days; and also after that, when the sons of God came in unto the daughters of men, and they bare children to them, the same became mighty men which were of old, men of renown.*
>
> GEN 7:4 *For yet seven days, and I will cause it to rain upon the earth forty days and forty nights; and every living substance that I have made will I destroy from off the face of the earth.* ...
> 11 *In the six hundredth year of Noah's life, in the second month, the seventeenth day of the month, the same day were all* **the fountains of the great deep broken up, and the windows of heaven were opened.** *12 And the rain was upon the earth forty days and forty nights.* ...
> 19 *And the waters prevailed exceedingly upon the earth; and all the high hills, that were under the whole heaven, were covered.* ...
> 8:1 *And God remembered Noah, and every living thing, and all the cattle that was with him in the ark: and God made a wind to pass over the earth, and the waters asswaged; 2* **The fountains also of the deep and the windows of heaven were stopped,** *and the rain from heaven was restrained;*

Some of the events seem to fit with Noah's flood in that we have the fallen *sons of God* and copious water from the *fountains of the great deep* and the *windows of heaven*. And perhaps some of the Greek myths of the *mighty men which were of old, men of renown* were not entirely myths. However, it is also clear that Noah's flood was not the first flood to cover the earth.

> GEN 1: 2 *And the earth was without form, and void; and darkness was upon the face of the deep. And the Spirit of God moved upon the* **face of the waters.** ...
> 9 *And God said, Let the waters under the heaven be gathered together unto one place, and* **let the dry land appear***: and it was so.*

The Apostle Peter wrote about a flood and the perishing *heavens* (plural heavens). Theologians have argued among themselves about which flood he described. Was it Noah's flood or the flood of Genesis 1:2?

2PE 3:4 *And saying, Where is the promise of his coming? for since the fathers fell asleep, all things continue as they were from the beginning of the creation. 5 For this they willingly are ignorant of,* **that by the word of God the heavens were of old, and the earth standing out of the water and in the water: 6 Whereby the world that then was, being overflowed with water, perished:**

Either scenario has profound implications. Did life exist—cities, birds, pre-Adamite man, fallen sons of God exist previous to the Genesis 1:2 flood? Or, was the first life (plants, animals, and man) something that occurred after Genesis 1:2? Whether **man**, in some genetic form, existed before the Genesis 1:2 flood can be argued, however, the scriptures do make it clear **that the sons of God did exist before** the foundation of the earth was laid: *Where wast thou when I laid the foundations of the earth ... When the morning stars sang together, and all the sons of God shouted for joy?*

Before we go further, let's refer back to our *Fiddler on the Roof* paradox when the father bemoaned, "If I try and bend that far, I'll break!" First, let us examine some of the early traditions of science that were ultimately broken and progress resulted. Hardly anyone would disagree that science has a trail of discarded theories as new information became available and insights are gained. Louis Pasteur brought a tremendous light to the dark ages of the practice of medicine, but he was ridiculed by his peers in the scientific realm. A fascinating account of this ridicule was found on a website of the National Health Museum.

> "From the time of the ancient Romans, through the Middle Ages, and until the late nineteenth century, it was generally accepted that some life forms arose spontaneously from nonliving matter. Such spontaneous generation appeared to occur primarily in decaying matter. For example, a seventeenth century recipe for the spontaneous production of mice required placing sweaty underwear and husks of wheat in an open-mouthed jar, then waiting for about 21 days, during which time it

> was alleged that the sweat from the underwear would penetrate the husks of wheat, changing them into mice. Although such a concept may seem laughable today, it is consistent with the other widely held cultural and religious beliefs of the time."

Ultimately, Pasteur's critical peers had to apologize. In the world of geology and biology, it is almost a given that the professors will ask students to check any religious beliefs about creation at the door. And any mention of God is often ridiculed as something based on myths. If a professor does not fully embrace every jot and tittle of Darwin's theory of evolution, many universities will not even consider their application (even if they are willing to embrace micro-evolution within a species). Many of the 'scientific' journals will reject any paper that hints of the hand of God in creation.

Now, let's look at the other side of the coin – where church 'tradition' finally reached a breaking point. Church leaders insisted that the earth was in the central position in spite of a lack of scriptures that said this to be the case – it was a tradition. But, eventually the evidence presented by Galileo and Copernicus prevailed that the earth rotated around the sun. Ultimately, the church of that location had to apologize.

The problem arises that both groups make statements that have untruths mixed within them. Since the statements have been made, they cannot be retracted because of pride—instead the untruths become fossilized. The fossils become idols—and the untruth in an idol cannot be bent to the truth because it might shatter the egos of those idolize it.

Within the Christian communities, it is no secret that disputes arise over various doctrines. For example, with the Hebrew society circumcision was traditional. However, Paul had quite a little spat with Peter over whether circumcision was required for Christians. Paul wrote: *But when Peter was come to Antioch, I withstood him to the face, because he was to be blamed. For before that certain came from James, he did eat with the Gentiles: but when they were come, he withdrew and separated himself, fearing them which were of the circumcision.*

Today, it is no secret that Christians have disputes over such things as 'the rapture' and whether Christians will be present during the 'tribulation' period upon the earth. Some have jokingly said that the battle of Armageddon will be fought by Christians arguing about details of rapture. It seems unlikely that one's basic salvation would be affected by the exact events of the blowing of *the last trump*, but it is likely one side or the other might be surprised at its unfolding. Another lesser point of contention has been how to interpret the creation process. This involves traditions within the Christian community and also traditions within the scientific community. Which traditions will be proven to be true — and which will bend until they finally break in light of truths co-existing in both science and scripture?

At some time in our lives, most of us have visited a barber shop with mirrors on both the front wall and back wall. As one looks at the images, a seemly infinite number of images fade off into the distance. This parallels the question that some scientists often ask theologians: Who created the heavens and earth? God. Who created God? God always was. What was God before he always was? … … … etc.(n+1)

On the other hand theologians often ask scientists: Where did life come from? It came from a primordial pool. Where did the primordial pool come from? From the big bang. Where did the big bang come from? … … … etc.(n+1)

It is clear that **both** theologians and scientists are trying to solve an infinite regression puzzle—the answers are still being sought and lie outside of the dimension that we call time. We probably should not try to repeal the law of gravity because it was 'discovered' by a Christian, Isaac Newton. (Interpretation: Newton might have believed in Intelligent Design—therefore his 'theory' is suspect!) On the other hand, theologians should realize that agnostics and atheists may also make valid scientific discoveries. *All things were made by him; and without him was not any thing made that was made. In him was life; and the life was the light of men. That was the true Light,* **which lighteth every man that cometh into the world.**

So, we go back to our basic question regarding the geologists and the paleologists and the framework of creation. They, no doubt, would be quite happy to find that time periods existed before the flood described in Genesis 1:2. This would correspond with geologic formations and various animal and plant fossils developed over significant periods of time.

Will lost civilizations be found in the channels of the sea as described by this scripture?

2SAM 22:16 *And* **the channels of the sea appeared, the foundations of the world were discovered,** *at the rebuking of the LORD, at the blast of the breath of his nostrils.*

If I Try And Bend That Far, I'll Break

It would seem that some mystery about the foundation of the earth exists in the channels of the sea. While there are many translations of the beginning verses in Genesis, one of the more literal translations (Concordant Literal) presents it this way (The bold words have direct correspondence to the Hebrew, while the normal type was filled in by the translator):

In a beginning Created by the **Alueim** were '**the heavens and** ' **the earth. Yet the earth became a chaos and vacant and darkness was on the surface of the submerged chaos.**

On the other hand, if man did not come on the earth until the creation of the 'sixth day' as described in Genesis 1:27, other groups will say, "I told you so! Why did you not believe? It's right there in the Bible!" *God created man in his own image ... And the evening and the morning were the sixth day.* But, at the same time, these groups (along with the scientists) may have some bending and breaking in order to deal with the scripture: *Neither do men put new wine into old wine skins: else the skins break, and the wine runneth out, and the skins perish: but they put new wine into new skins, and both are preserved.* New wine (discoveries) put into old cracked and flawed wine skins (traditions) will not have a good outcome.

But curiously enough, while science and theologians have been at odds for quite some time, it does appear that the more science discovers, the closer it coincides with scripture. It now appears that time and space is simply an accordion—it can be expanded and contracted between the book ends of the everlasting and the everlasting. One must wonder if the limitations of time and space apply to the spirit world and to the human spirit that resides within each of us. We end with another flashback to the period of Genesis written by David the psalmist:

PSALMS 90:1 *Lord, thou hast been* **our** *dwelling place in all generations.* **2 Before the mountains were brought forth, or ever thou hadst formed the earth and the world,** *even from* **everlasting to everlasting**, *thou art God.*

This is a concept that neither scientists or theologians have been able to fully grasp—but as the Apostle John wrote: ... **that there should be time no longer**: *But in the days of the voice of the seventh angel, when he shall begin to sound,* **the mystery of God should be finished**, *as he hath declared to his servants the prophets.*

And ye shall know the truth, and the truth shall make you free.

- Context Scripture Chapters: *Job 26, Rom 11, John 1&8&12 , Pro 8, Heb 11, Gen 1&3&6&7 , Psa 82, Jer 4, IIPet 3, Gal 2, IISam 22, Psa 90, Rev 10,*

Article Nine

MRI OF BRAIN SEEN BY HEBREW PROPHETS?

How Did They Know About Nervous System Anatomy?

King Solomon, while in an Ecclesiastical mood, once wrote: *The thing that hath been, it is that which shall be; and that which is done is that which shall be done*: **and there is no new thing under the sun.** *Is there any thing whereof it may be said, See, this is new? it hath been already of old time, which was before us.*

While humankind has invented marvelous new Magnetic Resonance Imaging machines that can peer deep within our body to inspect internal bone and tissues—is it possible that this knowledge was also available to the ancient Hebrew prophets, as recorded in the scriptures? Could these prophets have known about both the central nervous system and the spinal nervous system and posited this hidden wisdom in plain sight among other scriptures and also used parables to disguise the deeper truths? To answer this question, one must have both an understanding of the human nervous system and also be conversant with the scriptures. We will need to look at the 'discoveries' of modern day medicine and compare these to the knowledge of the ancients.

King David gave hints that certain information was hidden in parables when he wrote in the Psalms: *Give ear, O my people, to my law: incline your ears to the words of my mouth. I will open my mouth in a parable: I will utter dark sayings of old:* ... Is there a secret book which was the source of his knowledge — or was the knowledge just conveyed to him in parabolic symbols? He later wrote: *I will praise thee; for I am fearfully and wonderfully made: ... I was made in secret, and curiously wrought in the lowest parts of the earth. ... in thy* **book** *all my members were written, which in continuance were fashioned, when as yet there was none of them.*

The story of becoming aware of the biblical code about the human nervous system began when we were struggling with trying to understand a medical section newspaper article about the spinal cord nervous system being a twelve lane freeway of information flowing back and forth to the brain. This seemed to have some relationship to the 'twelve gates' described in Revelation and perhaps to the thirty-three spinal vertebra steps so often referenced by eastern and later western mystics. Unfortunately, the twelve lane analogy for the spinal cord did not at all fit—we found out that the spinal vertebra system was much inferior to something much higher and spiritually superior.

However, my most trusted nursing consultant said, "The article has it wrong. The twelve lane highway should not refer to the nervous system in the spinal cord; it should refer to the cranial nervous system. It has twelve nerves." Since she has had extensive experience assisting with brain surgery, it was time to follow-up on what was thought to be a 'break-through.' [It isn't rocket science engineering, but it **is** brain surgery—more complicated than any rocket's brain.]

The follow-up was this: The spinal cord nervous system **does not have twelve nerves**—it has **thirty-one nerves**! While the analogy of a freeway—with information flowing back and forth—may be crudely useful in helping a layman to understand the spinal cord, the twelve lane description is incorrect. It is the **cranial nervous system that has twelve nerves** and these nerves serve the cranium or the head. Could these be the twelve gates described in Revelation? But, then she said, "You know that these twelve nerves are paired?" How can that be? It was pointed out that we have two eyes, two ears, two nostrils, and so on. [This is why a stroke in a brain hemisphere can affect one side of the body and not the other.] But, this has greater significance when one considers that for the twelve paired nerves, two times twelve equals twenty-four. It looks like we have twenty-four senior [or elder] nerves. So?

All of a sudden, this light shone upon hidden information in the scriptures. As an appetizer, let us consider that the scriptures discuss that we are *a* **city** *that is set up on a hill, that great* **city***, the holy Jerusalem, descending out of heaven from God … And had a wall great and high, and had* **twelve gates***, … behold,* **the kingdom of God is within you***, … And round about* **the throne** *were four and twenty seats: and* **upon the seats I saw four and twenty elders** *sitting.…* Amazing! Twenty-four elders representing twelve nerve pairs! But this is only the tip of the iceberg. To see the rest, we will attempt to summarize what modern medicine 'knows' about the brain and nervous system and compare it to the hidden wisdom in the scriptures.

For those highly skilled in the anatomy of the brain and nervous system, please bear with us lay people as we consult a book entitled *ABC's of the Human Mind* [IBSN 0-89577-345-7] which was published by the Reader's Digest Association, Inc. On page 86, the two divisions of the nervous system are succinctly described.

"**Central Nervous System**, or **CNS** is composed of the brain and the spinal cord, the bundle of nerves running down the hollow of the bony spinal column.

Peripheral Nervous System, or **PNS**, is all the nerves branching out from the central nervous system and re-branching out to the farthest reaches of the body. Some of these nerves such as those serving the eyes and ears, connect directly to the brain (12 pairs) and are called *cranial nerves*. Those that branch out from the spinal cord to the rest of the body (31 pairs) are called *spinal nerves*."

There we have it, twelve nerves for the cranial nervous system and thirty-one for the spinal nervous system. Most of the cranial nerves sense events taking place in the head, but a few — like the vagus nerve — which is called the 'wanderer'—control parts of the body from the neck down. Jacob had twelve sons: ...*and at the gates twelve angels, and names written thereon, which are* **the names of the twelve tribes** *of the children of Israel:* Like Joseph of old, the vagus nerve—separate from the spinal cord—wanders down into the lower Egypt of our body and is primarily involved with controlling the heart—the 'viscera' part of our body. This is the place where we have a 'gut' feeling about something. The scriptures tell us that ***The spirit of man*** *is the candle of the LORD, searching all the inward parts of* ***the belly*** and *But let it be* ***the hidden man of the heart***, *in that which is not corruptible, even the ornament of a* ***meek and quiet spirit*** ... and ... ***Out of Egypt*** *have I called my son.*

Remember the scriptures about the city that *had* ***a wall great and high***, *and had* ***twelve gates*** *and* ... *there shall in no wise enter into it any thing that defileth* ... *and he placed at the east of the garden of Eden* **Cherubims**, *and* ***a flaming sword*** *which turned every way, to keep the way of* ***the tree of life***. "Oh," you say, "you are just spiritualizing everything." Perhaps, but maybe you are one that insists the Lamb and Lion in Revelation are actual animals? Then, again, consider that God is not limited to only one realm, be it the spiritual or the natural realm.

Is there a part of man's brain that has untapped potential that is not yet being used? Curiously, the brain has a high wall around it called the blood barrier. On page 82 of the ABC book we find this description:

"What is the blood-brain barrier?

In many ways, the brain is the most protected organ in the body. It even has a special arrangement with the blood stream.

Before it can reach the brain, a substance in the bloodstream must pass through a dense thicket of cells and capillaries that scientist call the blood-brain barrier. These cells and capillaries mesh in such a way that only certain molecules can pass: small oxygen molecules, for example, gain easy access."

But what about the 31 spinal cord nerves—what is their function. These nerves control most of the body from the head down. Are these in charge of our 'lower' or 'carnal' nature? Strangely enough, the brain has a threefold command structure just like the tabernacle in the wilderness and also Solomon's temple. *What? know ye not that your body is the temple ...* The 31 spinal nerves are the 'outer court' of our temple and strangely enough these nerves are brought into the cranium and expands into a swelled stem area called the 'olive'. This brain stem area is called the reptilian brain because it is the same configuration as the brain of cold blooded reptiles. When the spinal cord is shown in full relief, it looks like—of all things—a serpent. [One begins to have second thoughts about mysticism which is focused on raising 'spinal energies' until they conquer right through to attempt to subdue higher consciousness.] Is it just co-incidence that the top of the spinal cord is named the 'olive'? Does this scripture have any relationship to the 'olive' in our temple—or to a cocoon splitting open with the carcass of the *'old man'* being left behind in the transformation to a butterfly? *And his feet shall stand in that day upon the mount of Olives, which is before Jerusalem on the east, and the mount of Olives shall cleave in the midst thereof ...*

Oh, but there is more. The second level of the brain is the cerebellum. The cerebellum is sometimes called the animal brain. Animals have both a reptilian brain stem and a cerebellum. The cerebellum is concerned with **balance** and body movements. Can one strike a balance between good and evil? The cerebellum balances the information from the reptilian brain with the higher reasoning power of the holiest court of the brain, the overspreading wings of the cerebrum. You might say that we have Adam and Eve in the garden of our mind [or soul] having to make a decision between the lower nature of reptilian desires and the higher nature of the tree of life. *In the midst of the street of it, and on either side of the river, was there the tree of life, which bare twelve manner of fruits, And the foundations of the wall of the city were garnished with all manner of **precious stones** … And the twelve gates were twelve pearls:*

Unfortunately, both the serpent and man were banished from the garden containing the tree of life. Man's intellect was greatly diminished to only a fraction of his potential. *In the sweat of thy face shalt thou eat bread, till thou return unto the ground; for out of it wast thou taken: for dust thou art, and unto **dust** shalt thou return.* The serpent that once had the wisdom of the sacred stone nerve axion 'firing' gateways was banished to the dust of earthly spinal cord desires. *And the LORD God said unto the serpent … upon thy belly shalt thou go, and **dust** shalt thou eat all the days of thy life: By the multitude of thy merchandise they have filled the midst of thee with violence, and thou hast sinned: therefore **I will cast thee as profane** out of the mountain of God: and I will destroy thee, O **covering cherub**, from the midst of the **stones of fire**.*

But what is this mountain of God that is covered by the cherubs—is it not the third court of the temple—the holiest place? Is it not here where the cherubim's overspreading wings cover the ark? The cherubim are curiously described: *And the priests brought in the **ark** of the covenant of the LORD unto his place, to **the oracle** of the house, into the **most holy place**, even **under the wings of the cherubims:***

The cherubim have a unique configuration in the holiest place and so does the upper brain [the cerebrum] in the cranium. You see the two hemispheres of the cerebrum are not attached to the interior walls of the skull cavity and they overspread the pineal gland which is positioned between them. Similarly, *the cherubims within the inner house: and they stretched forth the wings of the cherubims, so that the **wing of the one touched the one wall, and the wing of the other cherub touched the other wall; and their wings touched one another in the midst of the house.***

Are *cherubims* a code word for the hemispheres of the cerebrum and the ark a symbol for the pineal gland—a good question? Does the pineal gland gather light whereas the scriptures say, *And the city had no need of the sun, neither of the moon, to shine in it: for the **glory of God did lighten it**, and **the Lamb is the light** thereof.*

However, this seems to be getting 'way out there.' Reptilian brain—you've got to be kidding? Does it **really, really** have any relationship to human anatomy? Let's go to page 62 of our ABC book to check it out.

"Why are some parts of the brain called reptilian and mammalian?

Neuroscientist Paul MacLean sees the human brain as really three brains. The first is made up largely of structures at the very top of the spinal stem. Because it resembles in some ways the total brain possessed by reptiles, this part of our brain is called reptilian in MacLean's terminology. Like the entire brain of a lizard or snake, it is mainly devoted to life support, such as the regulation breathing, heartbeat, and muscle movements, and to basic drives such as eating, mating, and self-protection.

While retaining generally the brain abilities of reptiles, mammals add dimensions to living. Their behavior, for example, involves complex emotional responses. They snarl, skulk, cringe, purr, growl, yelp in excitement or fear, nuzzle, teach their young, do tricks for rewards, wag their tails, show affection, and may even look ashamed or guilty when they have done something wrong. Such behavior is linked with a section of the mammalian brain that is not nearly so well developed in reptiles: the limbic system. For this reason, MacLean sees the limbic system as the core of our "second brain." If it does not function properly, he found in experiments with hamsters, the young lose interest in play and mothers stop their maternal behavior. The loss of limbic functions makes mammals behave more like reptiles.

Our third brain, according to MacLean, is composed of the outer bulges of the cerebrum and the overlying cerebral cortex. It is the reasoning brain."

Curiously, Solomon's temple had three levels and so does the brain. Around the outside of Solomon's temple were thirty chambers plus the main entry chamber which gives a total of thirty-one chambers. When King David received the plans for the three courts, for the cherubim, and the many chambers, he described how he got the information: *All this, said David, the LORD made me understand in writing by his hand upon me, even all the works of this pattern.* Was it just co-incidence that there were thirty-one chambers around the outside of Solomon's temple and there are thirty-one spinal cord nerves? If we relate the thirty-one spinal cord nerves as serving our lower nature, was it a co-incidence that when Joshua went into the promised land, he was told to subdue the kings of that land: *And these are the kings of the country which Joshua and the children of Israel smote on this side Jordan ... all the kings thirty and one.*

Here we get a clue as to how God moved on the Hebrew kings and prophets. They became oracles by the light within the ark of their pineal gland and from the well of the spirit rising up [out of the south] from the spirit through the vagus nerves. This river of consciousness is intuitive and right brain oriented. *He that believeth on me, as the scripture hath said, out of his belly shall flow rivers of living water.* Also, we read ... *and the waters came down from under from the right side of the house, at the south side of the altar.*

The prophets took down the information out of the spirit and transferred it across that vast interconnecting bridge between the two hemispheres called the corpus callosum to the left brain which translates spiritual concepts into words. Properly done, the word and the spirit will agree. Apparently, this is the pathway by which the Hebrew prophets of old were given an 'MRI scan' of the human nervous system and recorded it in the scriptures.

There are those reading this that are right brain oriented and they will know intuitively 'in their spirit' whether what is being written is true. There are others that are left brain oriented. They will need to reason everything out by reading it in the full context of the scriptures—excerpts of which have been quoted here. Both approaches can lead to a bridge between Word and Spirit for this 'earth' which is our human body: *In the beginning was the Word ... And the Spirit of God moved upon the face of the waters.*

While what we have presented here is only the tip of the iceberg of the patterns recorded in the Bible, there is so much more. Some of these codes are presented in a table on page 67 of the Chatan N. Kallah book, *The Dove Code* [ISBN 978-1-58169-299-0]. We join the dialog as Sarah is asking her university roommate, Kim, a question. The subject is the symbolism of the Mount of Olives being split in two as described by the prophet Zechariah.

****

""Right, and then where would the head of the serpent be?"

"I suppose it would be the spinal cord. Uh oh, here we go again!" exclaimed Kim. "I've seen pictures of the spinal cord in my anatomy book, and it does look like a serpent!"

"Yes, and the medical folks call the very top of it the 'olive' because it is swelled up like an olive."

Kim asked, "Do you mean we have a Mount of Olives right inside us? And, isn't there a scripture that when Jesus returns, the Mount of Olives will be split in two?"

Sarah confirmed the internal split of the Mount of Olives. But she admonished Kim to hold on to her spinal cord because she would need it until her terrestrial body was changed into a celestial body. She then revealed a flash of insight that the top of the spinal cord (the olive) is the altar where we must sacrifice the fleshly desires of the lower natures. She reiterated Ezekiel's insight that the spirit bubbles up and flows out of the right side of altar and into our right brain.

We need to get off the anatomy subject so that we can look at the booklet on the ancient scrolls. So, I tell you what, I will give you a copy of the secret temple code that I put together from Aunt Myra's writings for your future reference. Here it is:"

Temple Code

1. **Central Nervous System - 12 gates of Jerusalem - 12 paired nerves - 24 elders of the mind – located in head**
2. *Peripheral Nervous System - 31 paired nerves in spinal cord - serves body from neck downward - 31 kings smitten by Joshua in battle for promised land -*
3. *Spinal cord - serpent enters garden of the soul*
4. *Top of Spinal Cord - altar - head of serpent - olive - mount of Olives - Gethsemane*
5. *Pomegranates - cells of the brain*
6. *Jachin - right pillar in front of Solomon's temple - right eye*
7. *Boaz - left pillar in front of Solomon's temple - left eye*
8. *Three courts of temple - three levels of brain (reptilian, animal, and cerebrum)*

"Wow, and all this has been in the Bible, literally right under our very noses! Why haven't we seen this before?"

"It wasn't time for it to be revealed," replied Sarah. "Now, we could go on for hours and hours on the subject of the Bible and anatomy. Much of it is coded in the tent tabernacle Moses built in the wilderness,

"Wow!" said Kim, "There's a lot of code here to digest. I remember in our physiology book that humans have three levels in their brain. The lower brain is called the reptilian brain because it is the same as that of reptiles. The middle brain or cerebellum is called the animal brain because animals also have it. Only man has the cerebrum which is the highest and third level of the brain. And you said there were three divisions in the tabernacle and also in Solomon's temple. OK, OK, I'll put it away for another time. Where do we go from here?" [Editor's Note: *See Appendix E for Myra's Discussion of the Temple Code*]

****

Indeed, where do we go from here? This is a short writing, but it is hoped that it will stimulate readers to dig deep within their spirit and deep within the scriptures for further insight. It is time to find a *treasure hid in a field*—in the field that we call our very own being. In the next article, we will dig deep within— into the very DNA within the nucleus of our cells.

- Context Scripture Chapters: *Eccl, Psa 78&139, Mat 5, Rev 21&4, Pro 20, Mat 2, Gen 3, ICor 6, Zec 14, Rev 22, Eze 28, 2 Chron 5, IIKin 6, IChr 28, Jos 12, Gen 1, Mat 13*

Article Ten

UNSEEN DIMENSIONS LIFE SOURCE?

Where Does Life Come From?

What animates the inanimate and causes it to have a life force? Would an inanimate rock ever express love or have dreams? Love and dream scenarios are thought to be of an abstract nature — not something of a material nature that can directly be handled, seen and felt. Now, if one could take a rock and dissolve its components in water and from a primordial soup go through a progression of steps that resulted in a flesh and blood human being—that would be something that we could see, feel and evaluate in a material way—material from material.

In 1952, scientists were thrilled by the results of the Miller-Urey experiment. This experiment pre-supposed an early earth atmosphere of water, methane, ammonia and hydrogen. This mixture was sealed in a flask and lightning was simulated by an arcing between electrodes. After a period of time—within the flask were found amino acids—the structural units that are the building blocks for proteins. Five amino acids were found at the time, but in 2008, the archived solution was re-analyzed and 22 significant amino acids were found rather than just the five originally found. As we will later discover, the 22 number is extremely significant in our search for the source of life from the eleventh dimension. Why so? There are twenty amino acids that are directly encased within the universal genetic code. Then, two more amino acids are indirectly used in the generation of proteins. This gives a total of 22 naturally incorporated amino acids which are called *proteinogenic* because of their function in the protein building process. It appears that the chicken (protein) came before the egg (DNA) because protein has a complete complement of 22 building blocks. This is *Eureka!* information and we will find out why shortly. Is it possible that someone knew about the configuration of the amino acid—DNA structure over 3500 years ago? Can we find **an independent pattern of DNA transcription over 3500 years old** that existed before mankind was aware of modern chemistry and computer programming? The scriptures tell us that Moses was given the pattern for the candlestick for the Hebrew tabernacle in the wilderness while he was upon the mountain top.

What is this eleventh dimension? In earlier articles in this series, we noted that scientists now **really** believe there are seven 'unseen' dimensions that actually exist beyond the four familiar dimensions of length, width, depth and time. Seven unseen dimensions plus four seen dimensions equals eleven total dimensions. We have related the seven unseen dimensions and the twenty-two almond blossoms engraved in the candlestick as symbolic of the seven Spirits of God and the twenty-two letter Hebrew alphabet as given in Psalm 119. Note that there are four almond blossoms in the main stem. Curiously Psalm 119 is an acrostic with the first eight verses beginning with the letter Aleph, the next eight beginning with the letter Beth, and so on for each of the twenty-two letters in the Hebrew alphabet. (An example of a three line acrostic for '**A**' would be: (**A**pples are healthy. **A**pricots are delicious. **A**vocados are green.)

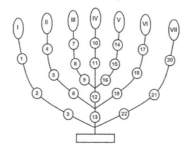

While the information about the synthesis of amino acids was quite exciting, in 1953 Crick and Watson developed the double helix model of DNA. This break-through answered many questions about the basic cell configuration but generated even more questions. The human genome was far more complicated than anyone had anticipated—so complicated that the effort to decode just one typical human cell extended into the next century. There are four nucleotide bases in this chain (adenine, thymine, guanine and cytosine, which are usually represented by A, T, G and C).

To help us visualize what is going on, let's give A, T, G and C some personality as **A**ndy, **T**iffany, **G**len and **C**arolyn. **A**ndy and **G**len are big, double muscled macho guys and **C**arolyn and **T**iffany are attractive,

The Articles of Configuration

petite gals. **A**ndy thinks **T**iffany is a knock-out doll and **G**len is sweet on **C**arolyn. Imagine a roller coaster car with two seats and each seat being able to hold a couple. Depending on who sat on the left side and who sat on the right side, there are eight possible sitting configurations for these couples. Now, with this visualization, we will switch to a diagram from the world of chemistry. Below, the subtitle and figure 3.8 is quoted from a book *Signature In The Cell* by Stephen C. Meyer [ISBN 978-0-06-147279-4]. The book compares the *random selection* role versus the *intelligent design* role as a probable source in life initiation processes.

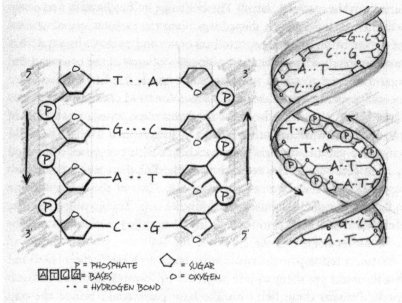

Figure 3.8. Antiparallel strands of DNA entwined around each other. Represented in two dimensions (*left*) and three dimensions (*right*).

Our very simplified explanation of the diagram: At the top left of the diagram in the roller coaster car, **T**iffany is holding hands with **A**ndy; then—one step down—we have **G**len holding hands with **C**arolyn. Then, **T**iffany likes the view from the right side better, so **T**iffany and **A**ndy switch sides. One step further down the ladder, **G**len and **C**arolyn switch sides. So, the A--T and G--C couples could sit in eight possible seating configurations in our roller coaster car – before it goes through the 'twister' loop. The eight seating positions are just like the eight verses in an acrostic for each of the 22 letters in the Hebrew alphabet. And many of the twenty-two letters in the Hebrew alphabet correspond to a part of the body—for example: *Gimmel* = neck = 3, *Yodh* = hand = 10, *Ayin* = eye = 70, *Resh* = head = 200. Also, as shown, each letter has a numeric value. Does this mean that genetic information is encoded in the acrostics of Psalm 119?

Maybe this field is worth searching for a pearl or a diamond. But on the other hand, not every field will contain a valuable gem. As Thomas Edison once said, "I speak without exaggeration when I say that I have constructed 3,000 different theories in connection with the electric light, each one of them reasonable and apparently likely to be true. Yet only in two cases did my experiments prove the truth of my theory." And also, he observed, "Many of life's failures are people who did not realize how close they were to success when they gave up."

Further puzzling scientists is how 'nature' managed to select only 'left-handed' amino acids for its building blocks. The problem would be similar to asking a human relations department to only hire identical twins—but also to only select left handed twins that are an exact mirror image of right handed twins. In other words, the twins are identical but not totally identical in that the amino acid building blocks actually used in proteins only bend light to the left. Oh, the moans coming from HR because of the tons of applications that they would need to sort through!

Why were we excited about having 22 amino acid building blocks in protein construction? Curiously, in the human genome, we receive 23 chromosomes from our father, and 23 chromosomes from our mother. These are joined together to make a total of 46 chromosomes in the cell nucleus. Why 23 instead of 22? The 23rd chromosomes determine whether we are male or female. Aside from male and female determination each of us receives 22 chromosomes from each parent. Suppose we persuaded Eve to jump back into Adam and become a rib—would Adam have had two sets of 22 chromosomes giving a total of 44 chromosomes in the cell nucleus? It is an interesting mental exercise to wonder if Adam was an androgynous being (both male and female) before the rib was removed.

For those of you who love to search for treasure, following is a field map that might possibly contain treasure(s). This is the genetic code for which its discoverers were given Nobel prizes in the fifties and sixties. Again we see the mysterious pattern of 22 building blocks. Also, if you care to count the three letter codes, you will find that there are (8 X 8 = 64) sixty four separate codes that the three position code readers (called codons) recognize. Perhaps unrelated, but the 22 letter Hebrew alphabet acrostic in Psalm 119 has eight verses for each Hebrew character.

1	Ala/A	GCU, GCC, GCA, GCG	12	Leu/L	UUA, UUG, CUU, CUC, CUA, CUG
2	Arg/R	CGU, CGC, CGA, CGG, AGA, AGG	13	Lys/K	AAA, AAG
3	Asn/N	AAU, AAC	14	Met/M	AUG
4	Asp/D	GAU, GAC	15	Phe/F	UUU, UUC
5	Cys/C	UGU, UGC	16	Pro/P	CCU, CCC, CCA, CCG
6	Gln/Q	CAA, CAG	17	Ser/S	UCU, UCC, UCA, UCG, AGU, AGC
7	Glu/E	GAA, GAG	18	Thr/T	ACU, ACC, ACA, ACG
8	Gly/G	GGU, GGC, GGA, GGG	19	Trp/W	UGG
9	His/H	CAU, CAC	20	Tyr/Y	UAU, UAC
10	Ile/I	AUU, AUC, AUA	21	Val/V	GUU, GUC, GUA, GUG
11	START	AUG	22	STOP	UAA, UGA, UAG

Most of you have seen representations of the DNA double helix with the little beads symmetrically placed on each DNA strand as it curves around the helix. The strand curves two and one-half times around each bead before proceeding on. These little bead spools on which the DNA strand is wound are called histones and are comprised of eight membered sub-units.

Early estimates placed the number of genes within the human chromosomes somewhere in the 40,000 to 100,000 range. Who knows where the final number will settle? However, if you like to dig in information treasure fields with your Google shovel, try using 22,000 human gene number as your search phrase. If you could only recruit a molecular biologist skilled in Hebrew linguistics, your archaeological team might make more rapid progress in digging out the artifacts—some of those ancient Hebrew alphabet characters being unearthed sure have funny looking shapes.

The scriptures allude to Jacob being a geneticist when he took *rods of green poplar, and of the hazel and chestnut tree; and pilled white strakes in them, and made the white appear which was in the rods. ... And the flocks conceived before the rods, and brought forth cattle ringstraked, speckled, and spotted.* Could the knowledge of the ancient Bible patriarchs have extended to the three codon readers that read the peeled open DNA during its transcription? Oh, come on! — how likely is that?

Intelligent design enthusiasts have pointed to the repeating patterns in the genetic code as a creative plan. However, if a pattern of this creative programming were discovered in the Psalms—written long before DNA was *discovered*—that would be quite startling.

But a number of scientists scoff at this Adam and Eve stuff and are much more interested in the *physical* world. After all *physicists* who study *physics* want everything to be proven logically and they shy away from 'myths'. Things need to be *seen* and reasoned out. But, scientists have also become comfortable with some parts of the unseen world as one of the seven unseen dimensions is said to be electromagnetism. It would seem that electromagnetism also has a spiritual dimension as described by the ancient Hebrew prophet Job: *Then a spirit passed before my face; the hair of my flesh stood up:* Scientists admit that they do not know much about the unseen, curled up dimensions. But, if they can accept electromagnetism as one of these dimensions—why all the hesitation about the possibility that the other six are also unseen dimensions of the spiritual realm? When it comes to going through the vail into the spiritual realm, is *physically* oriented mankind like a fish out of water? Why not consider the possibilities of what is behind the curtain?

Now faith is the substance of things hoped for, **the evidence of things not seen**.

... *Through faith we understand that the worlds were framed by the word of God, so that* **things which are seen were not made of things which do appear**.

What is behind the curtain in the spiritual world that causes the seen things in the physical world to appear as they do? Can we get life out of an inanimate rock dissolved in a primordial soup, or are there some unseen components that are absolutely necessary? Would the rock be barren as far as producing life?

What components might be necessary to conceive life? Perhaps a clue is given in the following scripture: *Through faith also Sara herself received strength to conceive seed, and was delivered of a child when she was past age, because she judged him faithful who had promised.* This seems so 'far out' when it comes to describing faith as the vehicle for life—but is it really any 'further out' than scientists telling us there is an unseen world of seven dimensions?

The Articles of Configuration

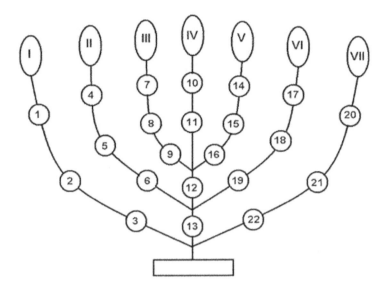

    Consider the main shaft of the candlestick above—it has four almond blossoms (shown as numbers 10, 11, 12, and 13). There is a scripture about the vine and the branches — without the main vine—the branches can do nothing. Genesis speaks of a garden with a river flowing out of it. Since scriptures are sometimes expressed in parables, we will not limit ourselves to a geographical Garden of Eden but will also consider that this garden may be symbolic of something else—our soul. *And the LORD shall guide thee continually, and* **satisfy thy soul** *in drought, and make fat thy bones: and* **thou shalt be like a watered garden**, *and like a spring of water, whose waters fail not.*

    Now let us visit the Garden of Eden and see what is flowing out of it. *And a river went out of Eden to water the garden; and from thence it was parted, and became into four heads.* Brain researchers have identified four different types of brain waves — which water the mind or soul. These are as follows:

    Delta waves (below 4 hz) deep sleep
    Theta waves (4-7 hz) sleep and deep relaxation
    Alpha waves (8-13 hz) relaxed and creative
    Beta waves (13-38 hz) when thinking, problem-solving, etc.

    What is the source of this river of consciousness? There is a scripture that identifies the 'belly' as a river source. *He that believeth on me, as the* scripture hath said, *out of his belly shall flow rivers of living water*. But, what is located in the belly that can flow like a river to the mind and part into four wave patterns? King Solomon identified it as a candle located in the belly—you might say that is where we have our 'gut feeling': *The spirit of man is the candle of the LORD, searching all the inward parts of the belly.* So, we have traced the river back to the spiritual, unseen flame of the main shaft candle which lights part of the seven dimensioned candlestick. Is this the source of life flowing from the seven 'curled up' dimensions now being investigated by our scientists?

Consider the reports of those who have had 'near death' experiences. Some have reported numbness or tingling as the 'body electric' recedes from the extremities and then finally leaves the body. Then it seems when the spirit passes into an unseen dimension, it is there but those in the physical world can't see it. A common report from near death experiences is that of rising from the body and watching the desperate efforts of those around the operating table trying to resuscitate them. And then, many have reported going through a tunnel of light into perhaps a parallel universe where unusual experiences await them.

The physical world requires tremendous sources of energy fed into large hadron colliders (like those at Cern) to attempt to go beyond the vail—however the human spirit seems to be able to do it effortlessly.

Scientists shake their head in wonder at the concept of seven, unseen, curled up dimensions. Much of this exploration is done mathematically and the equations can be horribly complex, but from time to time discoveries are made that simplify complex problems into simple and elegant solutions. There is a documentary about a young adult who was able to do extremely complex numerical computations in his head within just a few seconds—sort of like asking what the square root is of 3,794,704 and getting an answer back in seconds. A neuroscientist investigated the phenomena and found that the computations were not being done in the 'normal' math area 44 of the brain. Instead they were centered in the parietal area controlling vision. Perhaps, this man saw into the 'unseen' dimensions—not seen by normal human beings.

But what may be extremely complex to our scientists today may be child's play for those who can 'see' the unseen. *My kingdom is not of this world:*

Whosoever therefore shall humble himself as this little child, the same is greatest in the kingdom of heaven.

- Context Scripture Chapters: *Psa 119, Heb 11, Isa 58, Gen 2 & 30, John 7, Pro 20 John 18, Mat 18*

Article Eleven

THE ARTICLES OF CONFIGURATION

The Genesis Project
A Summary of Eleven Articles

These eleven articles are presenting various aspects of basic variables intimately involved in the configuration of the universe. We have been traveling through the forest and looking at the individual trees. Now it is time to fly over the forest and summarize and list the 'takeaway' themes in each article.

1. Genesis Chain of Custody
2. Earth's Magnetic Shield in the Bible?
3. Are You Smarter Than a PhD Physicist?
4. Light at the Top of the Mountain
5. Expanding Universe in Scriptures?
6. Time Travel (One Day = 1000 Years)
7. Do Life Forms Exist Among the Stars?
8. If I Bend That Far, I'll Break!
9. MRI of Brain Seen by Hebrew Prophets
10. Unseen Dimensions Life Source?
11. The Articles of Configuration Summary

Genesis Chain of Custody: Before the early processes listed in the first part of Genesis, the raw material for formation came from the *highest part of the dust of the world*. This dust existed before the world was—therefore its age is independent of when the earth was formed into a spherical planet.

The fourth day concerning the rule of the sun, moon and stars in the heavens describes the earth when *the cloud the garment thereof, and thick darkness a swaddlingband for it* (the earth) was cleared away. The rule of the sun, moon and stars *signs, and for seasons, and for days, and years* then became apparent.

Earth's Magnetic Shield in the Bible? While scientists in recent centuries have 'discovered' the magnetic shield around the earth – this concept was mentioned in these ancient scriptures:

For the LORD God is a sun and shield ...

... for the shields of the earth belong unto God ...

Without the sun, man could not exist. Without the magnetic shield around the earth, mankind would be decimated by solar wind radiation.

Are You Smarter Than a PhD Physicist? This article explores the configuration of the seven lamped candlestick that was present in both Moses' tabernacle in the wilderness and later in Solomon's temple. Scientists have been fervently at work developing string theory and membrane and their work leads to some unusual conclusions. They theorize that there are a total of eleven dimensions that make up the universe. Four of these dimensions are the readily observed dimensions of length, width, depth and time. Seven are thought to be 'curled up' unseen dimensions that really exist but scientists are somewhat at a loss to describe.

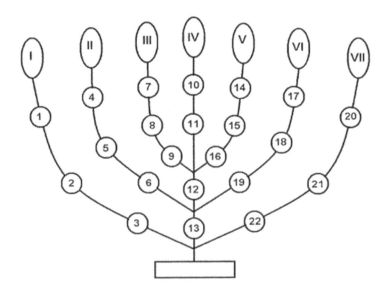

The exploration of membrane theory plus string theory results in mathematical equations with eleven and twenty-six parameters. When one subtracts out the four seen parameters, the unseen parameters are seven and twenty-two. The seven lamped candlestick has twenty-two almond blossoms in its configuration (See the adjacent diagram of the candlestick). This is the same number of letters in the Hebrew alphabet which is given in Psalm 119. Hebrew scholars think that this is the language of creation. The almond rod which blossomed belonged to Aaron, the brother of Moses. This was the rod that Moses stretched over the sea when being pursued by the Egyptians. Many of the building blocks of nature number twenty-two and these are summarized at the very end of this article.

Light at the Top of the Mountain: This article chronicles the journey of string theorists and super gravity theorists as they struggled to come up with a Theory of Everything. The string theorists were like the branches and the super gravity had the vine. For a while the groups warred against each other over the number of dimensions in the universe—was it ten or eleven? The result was eleven which included seven unseen dimensions. Theologians can relate the seven unseen dimensions to the seven lamps which are called the Seven Spirits of God.

The article further chronicles the path to M-theory—or membrane theory. Membrane theory leads to the concepts of parallel or multiple universes. The scriptures do refer to *the third heaven* and *the lowest hell*.

Expanding Universe in Scriptures? Scientists have developed the concept of stretchable membranes with vibrating strings attached to them which make up the universe. The scriptures of antiquity contain numerous references to *God the LORD, he that created the heavens, and stretched them out; he that spread forth the earth*. Giants of science such as Einstein and Hoyle thought that the universe was neither expanding nor contracting—but existed as a steady state. The work of Hubble and many others showed that the universe was *stretched* out and is still expanding.

Time Travel (One Day = 1000 Years): It is written that *...one day is with the Lord as a thousand years.* What is the relationship of this scripture to time travel? Strangely enough when the contraction of time from 1000 years to one day is considered, one is approaching the breaking of the light barrier of 186,000 miles per second.

Further, the classic time travel example of one boarding a very fast rocket ship and having the aging process slowing down is examined. If the time traveler's earth bound contemporaries age much faster, then time elapsed in the earth realm is much greater than time relatively elapsed in the rocket ship realm. So, is age something different depending on which side of the looking glass worm hole one is positioned?

Do Life Forms Exist Among the Stars? Scientists have postulated that given the billions upon billions of stars and their planets, a relatively high probability exists that there may be life forms in other parts of the universe. And also, a probability exists that these life forms may have civilizations much more advanced than the earth. This article asks a very basic question: Could these more advanced life forms and civilizations be God and his Holy angels?

If I Bend That Far, I'll Break! The theme of this article can be summarized: *Neither do men put new wine into old wine skins: else the skins break, and the wine runneth out, and the skins perish: but they put new wine into new skins, and both are preserved.* Over the years, both scientists and theologians have found it necessary to shed the old wine skins of flawed ideas and traditions in order to embrace truth. And yet, both are finding truths that are embedded in the scriptures written from antiquity. This article further examines the Genesis scenario and the pre-Genesis scenario concerning the flood and *the higher part of the dust of the world.*

MRI of Brain Seen by Hebrew Prophets: This article shifts gears from the general configuration of the universe to human anatomy. *Ye are the light of the world. A city that is set on an hill cannot be hid.* The twelve gates of the New Jerusalem and its 24 elders are compared to the twelve **paired** cranial nerves of the central nervous system. The peripheral nervous system of 31 nerves (spinal cord) is compared to the 31 kings that were to be subdued in the Promised Land and also the 31 outer chambers in Solomon's temple. The three levels of human brain function (cerebrum, cerebellum, and reptilian) are compared to the holiest place, the sanctuary, and the outer court of the tabernacle. The two cherubim (with their wings touching the wall and touching in the middle) are compared to the two hemispheres of the human brain with the internal connector and pineal gland being related to the ark.

The Articles Of Configuration

Unseen Dimension Life Source? This article describes the search for the source of life. For too long scientists have been trying to find the source of life in the physically seen dimensions. Now that they have introduced themselves to the seven unseen dimensions, it is time to consider what they have 'discovered'. Many of the configurations of life seem to be symbolized in the seven lamps and twenty-two almond blossoms of the Hebrew candlestick.

Summary of Candlestick Building Block Symbols:

- **Seven Unseen Dimensions = Seven Lamped Candlestick**
- **Twenty-two Almond Blossoms = 22 Letters of Hebrew Alphabet**
- **Twenty-two Letters = Twenty-two Phyla of Life Forms**
- **Twenty-two Letters = Twenty-two Proteinogenic Building Blocks**
- **Twice Twenty-two Letters = Forty-four Phonemes in Language**
- **Twice Twenty-two Letters = Forty-four Human Chromosomes (Less the 23rd Male and 23rd Female Gender Chromosomes)**
- **The Sun Reverses Its Magnetic Field Every 22 years**

It is the glory of God to conceal a thing:
but the honour of kings is to search out a matter.

The Articles of Configuration are but chronicles of stepping stones leading to the doorway of comprehension of the constitution of the seen and unseen worlds. To take this path, we have had to discard some old wine skins of ideas that have become cracked and flawed. But, at the same time, old truths have had to be 'rediscovered'. This is summarized in the scripture:

Then said he unto them, Therefore every scribe which is instructed unto the kingdom of heaven is like unto a man that is an householder, which bringeth forth out of his treasure **things new and old**.

- Context Scripture Chapters: *Pro 8, Job 38, Gen 1, Psa 84 &47, IICor 12, Psa 86, Isa 42, IIPet 3, Mat 9, Mat 5, Pro 25, Mat 13*

ABOUT THE AUTHOR

Carl Armstrong is a native of Missouri and an engineering graduate of Missouri School of Mines and Metallurgy. He, and his wife Connie, have five children and have lived in Texas, North Carolina, and Tennessee. They currently reside in southeast Missouri on the farm where Carl spent his childhood. They have restored their 1795 farm home which has an interior log cabin.

He has worked with a number of companies producing medicines, organics, and farm products. He is president of a land corporation. Carl has a strong interest in both scientific and spiritual studies and has gone on Christian missionary trips to Haiti and Africa. He has also traveled to Israel to gain an appreciation of the context of the scriptures.

Log on to
http://AlphaAncientOmega.com/
for any information updates